Superfine Particle Technology

Noboru Ichinose, Yoshiharu Ozaki
and Seiichirō Kashū

Translated by Malcolm James, BSc,
for M.J. Shields, FIInfSc, MITI (Associates)

Superfine Particle Technology

With 145 Figures

Springer-Verlag
London Berlin Heidelberg New York
Paris Tokyo Hong Kong
Barcelona Budapest

Noboru Ichinose, PhD
Professor, Waseda University School of Science and Engineering,
3-4-1 Ohkubo, Shinjuku, Tokyo 169, Japan

Yoshiharu Ozaki
Seikei University, 3-3-1 Kitamachi, Kichijoji, Musashinoshi, Tokyo 180,
Japan

Seiichirō Kashū
Vacuum Metallurgical Co., Ltd., 5-9-7 Tokodai, Tsukubashi, Ibaragi
300-26, Japan

Translated by: Malcolm James, BSc, for M.J. Shields, FIInfSc,
MITI (Associates)

ISBN-13:978-1-4471-1810-7   e-ISBN-13:978-1-4471-1808-4
DOI: 10.1007/978-1-4471-1808-4

British Library Cataloguing in Publication Data
Ichinose, Noboru *1935–*
Superfine particle technology.
1. Particle technology
I. Title II. Ozaki, Yoshiharu III. Kashū, Seiichirō
620.43
ISBN-13:978-1-4471-1810-7

Library of Congress Cataloging-in-Publication Data
Ichinose, Noboru
Superfine particle technology/Noboru Ichinose, Yoshiharu Ozaki, Seiichirō Kashū
p. cm.   Translated from the Japanese.
ISBN-13:978-1-4471-1810-7 (U.S.)
1. Particles. I. Ozaki, Yoshiharu, 1942– . II. Kashū, Seiichirō. III. Title.
TA418.78.125   1991
620'.43—dc20   90-42962
                                                       CIP

Original Japanese edition published as *Cho Biryushi Gijutsu Nyumon* by Noboru
Ichinose, Yoshiharu Ozaki and Seiichirō Kashū
© 1988 by Noboru Ichinose, Yoshiharu Ozaki and Seiichirō Kashū, published
by Ohmsha, Ltd, Tokyo, Japan
English edition © 1992 Springer-Verlag London Limited
Softcover reprint of the hardcover 1st edition 1992

Translation rights arranged with Ohmsha, Ltd.

Typeset by Photo·graphics, Honiton, Devon
2128/3830-543210   Printed on acid-free paper

Preface to the Japanese Edition

If a substance is repeatedly subdivided, the result is what are known as "microscopic particles". These particles are distinguished from the solid mass which they originally formed by the size of the surface area per unit weight. This simple difference holds true down to a certain lower size limit, and when this limit is exceeded, a new state of matter is reached, in which the behavior of the particles is quite different to that of the original solid. Particles in this state are termed "superfine particles", and are distinct from ordinary particles.

The size of the superfine particles, that is to say the size limit below which particle behavior is completely different from the behavior of the original solid, varies a good deal depending on the physical properties of the substance in question. Properties such as magnetism and electrical resistance are closely related to the internal structural properties of the particles themselves, such as the magnetization processes of their respective magnetic domains, and the mean free path of charged bodies. This internal structure therefore limits the size of the superfine particles. In ceramic processing, on the other hand, the surface area of the particles themselves becomes an even more important factor than their internal structure. In this case, the size of the superfine particles is determined by the interaction between water and solvents on the surface of the particles.

In this book, a superfine particle is defined as a particle with a diameter in the range 1 to 100 nm. In Chapters 1 to 3 we describe the basic nature, the physics and the chemistry of superfine particles, and in Chapters 4 and 5 we discuss physical and chemical production methods. In Chapter 6 we deal with engineering applications of superfine particles, and describe applications in electronic, magnetic, optical, sintering, catalyst and sensor materials.

Books on the general theory of superfine particles have already been published, as have books on particular aspects, and explanatory texts, but we feel that ours is the first book to give an overall view of their basic nature, material properties, production methods and applications. If our readers find this book of some use, however small, in the development of superfine particles, we shall be entirely satisfied.

Finally, we would like to express our sincere gratitude to everyone involved in the publication of this book, and in particular to the staff of the Publishing Department at Ohmsha.

June 1988   *Noboru Ichinose, on behalf of all the authors*

Contents

Chapter 1

Fundamentals of Superfine Particles

Fine particles with a diameter of anything from 1 to 100 nm are commonly known as superfine particles. Superfine particles may, however, constitute a completely new form of matter which exhibits patterns of behavior quite different from those of the substance's corresponding solid form.

In this chapter, by way of an introduction to the fundamentals of superfine particles, we will first of all outline their general properties. This outline will be followed by a section dealing with the volume effect of superfine particles as typified by the Kubo effect. We will then touch on the surface effect whereby the number of atoms forming the surface of a superfine particle is proportionately greater than the number forming the surface of the corresponding bulk material. Finally there is a section dealing with the interactions taking place either between one particle and another or between a particle and its interface area.

1.1 Introduction

If a substance is continually reduced in size it will eventually reach what is known as the superfine particle stage. Such particles can be distinguished from their corresponding bulk solid form by the size of their surface areas in relation to their weight. When this simple difference exceeds a particular value then a completely new form of matter is created which behaves quite differently from its corresponding bulk solid form. Particles which exhibit this type of condition are called "superfine particles" to distinguish them from ordinary particles.

The size at which a particle becomes a superfine particle, or in other words the size at which its behavior becomes completely different from that of its corresponding bulk solid form, varies considerably depending on the particular characteristics which are being observed. The properties of magnetism and resistance are intimately bound up with certain aspects of the internal structure of the particles themselves, namely the magnetization of the magnetic domain and the mean free movement of the carriers of electrical charges, respectively. The size of a "superfine particle" is in this way limited by its internal structure.

The "superfine particle" research project which formed part of the Japanese national program for the advancement of original scientific techniques [1] has resulted in the publication of a substantial number of findings, most of which relate to particulate metals and metallic compounds. The project itself has now been wound down. Recently, however, a considerable amount of attention has been focused on substances such as oxides and nitrides. In the field of ceramics, within which both oxides and nitrides may be included, the most important factor from the processing point of view is the size of a particle's surface area rather than its internal structure. For processing purposes, therefore, the critical size at which a particle becomes "superfine" is determined by the interaction of a particle's surface with, for example, water or some other solvent.

The phrase "superfine particle" itself may well conjure up an image of some amazing new contribution to the world of science from the realms of new materials technology. In point of fact, however, common clay which man has been manipulating skillfully for centuries in the creation of a multitude of different types of artifact is just one form of "superfine particle" which occurs naturally. It is likely, therefore, that the practical application of "superfine particles" with some kind of man-made functionality will also be possible using similar processing techniques to those applied in the case of naturally occurring types. The rest of this chapter will be devoted to a description of some of the fundamental characteristics of "superfine particles".

1.2 Properties of Superfine Particles

In describing the properties of substances we are normally referring to entities which consist of almost limitless numbers (up to 10^{23} or so) of individual atoms. The properties of a single atom and those of a substance which is formed from an agglomeration of such atoms are different, however [2]. Iron is highly magnetic whereas an iron atom is not. Similarly, of the various compounds formed with iron some are magnetic and some are not. It is therefore conceivable that it is not only these two extremes, as represented by a substance consisting of an infinite number of atoms and by a single constituent atom of that substance, that have differing properties but also that an intermediate type of matter consisting of a finite number of these same atoms may also have differing properties depending on the actual number of atoms in the substance in question.

Particles with diameters between 0.1 and 1.0 μm are known as fine particles and normally consist of between 10^9 and 10^{10} atoms. With particles of this sort of size and number of atoms there seems little reason not to attribute to them the properties originally attributed to substances thought to be constituted of infinite numbers of atoms. The situation with smaller particles is not quite so clear cut but it seems likely that there is a boundary beyond which their properties differ markedly from those of larger particles. These property differences stem from differences in volume between the substances and are thus known as the volume effect. In practice, however, it is not an easy matter to confirm the volume effect empirically. This is due to an additional effect called the surface effect which also starts to manifest itself as particle size is reduced.

Atoms on the surface of a solid encounter a different sort of environment than those deeper inside the solid, in that those inside the solid are surrounded by other similar atoms whereas the surface atoms have similar atoms on one side only. To this extent the surface atoms may be considered to have properties that are different from that of the interior atoms. However, since the number of surface atoms is extremely small by comparison with the number of interior atoms this property difference is of little practical significance except in cases where the role of the surface atoms assumes particular importance, such as in the reflection of light or in chemical reactions.

Generally speaking, a spherical atom with radius r has a surface area S which is equal to $4\pi r^2$ and a volume V which is equal to $(4\pi/3)r^3$. The proportional surface area σ of a particle can thus be expressed by the equation

$$\sigma \equiv \frac{S}{V} = \frac{3}{r} \propto \frac{1}{r} \tag{1.1}$$

If the particle is not spherical then the value of σ will be greater. The proportional surface area σ_F of a spherical particle with a radius of 1 μm (in other words with a diameter of 2 μm) is given by the equation

$$\sigma_F \geq \frac{1}{10^{-4}} = 10^4 \ [\mathrm{cm^{-1}}]$$

In other words, 1 cm³ of particles with diameters of 2 μm would have a surface area of at least 1 m². The huge size of this value may be more clearly grasped if it is compared with a regular 1 cm³ cube which has a surface area of 6 cm². One cubic centimeter of spherical superfine particles with diameters of 10 nm would have a huge surface area of 10 m².

If we now consider an atom as a cube with a side d then the number of atoms in a particle with a volume V is given by V/d^3 and the number of surface atoms is given by the ratio S/d^2. The total number of atoms Σ in the proportionate surface area is thus given by the formula

$$\Sigma \simeq \frac{S/d^2}{V/d^3} = \sigma_d \tag{1.2}$$

For a spherical particle with radius r this gives

$$\Sigma_b = \sigma_d = \frac{d}{r} = \frac{1}{r/d} \tag{1.3}$$

r/d is inversely proportional to the one-dimensional size of the particle as measured in terms of the number of atoms. For example, in the case of a spherical particle with a radius of 1 nm and a particle spacing of 0.2 nm, $\Sigma \doteq 0.2$. In other words 20% of all the atoms are surface atoms. On the other hand, in a similar spherical particle with a radius of 1 μm, $\Sigma = 2 \times 10^{-4}$. In other words surface atoms account for just 0.01% of the total. In particles with radii of 10 μm or more Σ is even smaller and in a normal crystalline particle the number of surface atoms is insignificant. Thus, where the diameter of a particle is large by comparison with that of its constituent atoms then the surface atoms can safely be ignored but when the particle diameter approaches more nearly that of its constituent atoms then the proportion of surface atoms must be taken into account. One of the most notable features of superfine particles is the significantly high proportion of surface atoms which they contain. This is otherwise known as the surface effect.

In the above example we considered a sphere with the smallest possible Σ but in an actual polyhedron or other solid figure with uneven surfaces the value of Σ will, of course, be larger. Under experimental conditions the number of surface atoms in an Fe cube of 2 nm has been measured at approximately 80%. Looked at from a different point of view we can say that since the value of Σ is determined by the ratio of particle spacing to particle size (r/d) as seen in Eq. (1.3) then the bigger the value of Σ the smaller the value of r/d. A small number of atoms lying along the radius of a particle means not only that there will be a relatively high proportion of surface atoms but also that their influence may extend further into the interior of the particle. It may be appropriate at this point to stop thinking in terms of the surface and interior of a superfine particle and to think instead in terms of a single molecule.

Particles which exhibit either one or both of the volume and surface effects are known as "superfine particles". Furthermore, the critical size of a superfine particle varies depending on the substance in question or the particular phenomenon which is being observed. The identification of that critical boundary and the description of what is happening within its confines is the task of superfine particle research and as such this constitutes an entirely new branch of physics, separate from atomic physics and the physics of condensed matter.

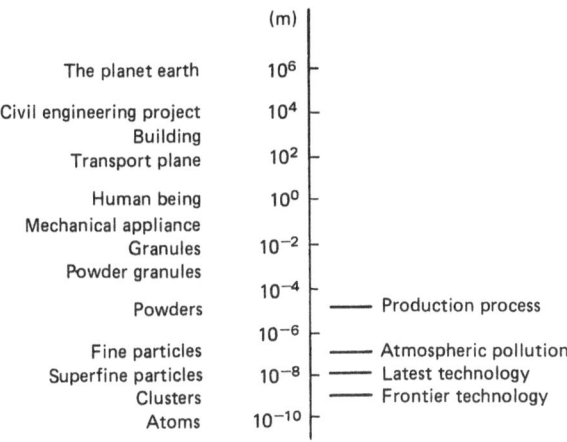

Fig. 1.1 Comparative size of bodies and fine particles.

We will now consider the size of a superfine particle. Figure 1.1 takes the human body as its starting point and indicates the relative size of a number of different physical bodies [3]. The upper end of the scale includes a number of bodies which outrank the human body in size, ranging from transport planes, buildings and civil engineering projects right through to celestial bodies. At the lower end of the scale come mechanical appliances, mechanical components and then the material that makes up such bodies, ranging from granules and powders down to fine particles, superfine particles and finally atoms. It is clear from this scale with man at the center that a superfine particle is more remote from a human being in terms of size than is the earth on which we live.

Table 1.1 shows the diameters and physical characteristics of particles along with a qualitative evaluation of the contribution made to their overall characteristics by the relative size of their surface area [4]. It is clear that quantum effects such as the volume effect are phenomena

Table 1.1 Diameters and characteristics of particles

Classification	Diameter	Atom count	Surface contribution	Characteristics
A	$> 1\ \mu m$	$> 10^{11}$	Bulk	Bulk
B	$1\ \mu m \sim 100\ nm$	10^8	Surface effect manifested	Bulk
C	$100 \sim 10\ nm$	10^5	Large surface effect	Bulk
D	$10 \sim 1\ nm$	10^3	?	Quantum effect
E	$< 1\ nm$	$< 10^2$	Bulk/surface effects disappear	Clusters, molecules

which manifest themselves on the boundary between substances in their macroscopic (bulk) form and in their molecular form. It was observed above that superfine particles could be defined as those particles which exhibit either the volume effect, the surface effect or both, but on a broader view we could safely define a superfine particle as any particle with a diameter d which falls within a range $1\,nm \leqslant d \leqslant 100\,nm$, where $1\,nm = 10^{-9}\,m = 10^{-3}\,\mu m = 10\,\text{Å}$ [5].

According to Hayashi [6] the main characteristics of superfine particles and their likely applications are as follows:

1. Large surface area per gram weight ($\geqslant$ several $10m^2/g$): heat or material exchange membranes.

2. Thin, uniform surface layer (film, for example a 2–5 atom layer of an oxide): auxiliary combustion agent for rockets (at the critical temperature all particles react immediately).

3. Stepped surface at the atomic level: high performance, high speed catalysts (reaction facilitated by large number of active points).

4. Can form an even distribution or mix in gases, liquids or solids: dispersant reinforcers, equalizing agents, aerosols.

5. Superfine particle chaining (diameter $\leqslant 30\,nm$, length/diameter $\doteqdot$ 2–100, fine, neck-like joints): magnetic recording media, molecular filters, electromagnetic wave absorbers, filters, etc.

6. The electronic energy level inside the particles is dispersed (Kubo effect): very low temperature, far infrared materials.

7. Standing wave lengths limited by size of particle: electromagnetic wave resonators.

8. Particle diameter determines level of mean free movement of electrons in solid: particular conductive materials, photoelectric conversion.

9. Easily carried, infiltrated, rejected or introduced anywhere inside an organism: physical treatments and medicines.

10. Superfine magnetic particles can be created inside certain types of bacteria: earth sciences and sciences dealing with the manifestation of biological functions.

11. Clouds of superfine particles consisting of elements such as C, Si, Fe and O round certain types of celestial body: study of stars, planets and evolution.

Particles in which the atoms can be counted (2–300) are known as clusters. Superfine particles in the widest sense may also include such clusters.

1.3 Volume Effect in Superfine Particles

The Kubo effect [7] which first pointed the way to superfine particle research is a typical example of the volume effect. When a large number of the valence electrons in a metallic atom come together they appear to form a continuous band of energy made up of conduction electrons, but on closer inspection it becomes clear that this continuous band of energy is in fact subdivided into energy levels equal in number to the number of electrons present. Where the number of electrons is very large the gaps between the individual energy levels is extremely small and this accounts for the appearance of continuity in this "band" of energy. Where, however, the number of atoms (i.e. the number of valence electrons) is small then the gaps between the energy levels are comparatively large and the energy levels themselves will appear discrete rather than part of a continuous band. It is thus more appropriate to regard the energy levels as discrete when the gaps between them are large, rather than when they become misaligned as a result of thermal vibration.

We will now draw together the principal features of the Kubo effect. The effect takes as its starting point the naive concept of the free electron. Normally a number N of free electrons enclosed in a cubic area of a certain size, say L^3, is treated as a system of metallic conduction electrons. Each of the atoms is in the stationary quantum state permitted within this area. If the wave function which constitutes the boundary condition determining this quantum state is assumed to be zero at the boundary then the quantum state is given by

$$\psi_{n_1 n_2 n_3} = C \sin \frac{n_1 \pi x}{L} \sin \frac{n_2 \pi y}{L} \sin \frac{n_3 \pi z}{L}$$

$$n_1, n_2, n_3 = 0, 1, 2, \cdots$$

(1.4)

and its energy is given by

$$\epsilon(n_1, n_2, n_3) = \frac{\hbar^2 \pi^2}{2mL^2} (n_1^2 + n_2^2 + n_3^2)$$

(1.5)

Each energy level is made up of a combination of two levels, one with a positive spin and one with a negative spin. If there is an external magnetic field H then the contraction can be expressed as

$$\epsilon(n, \sigma) = \epsilon(n) - \mu_B \sigma \qquad \sigma = \pm 1$$

(1.6)

Here μ_B is a Bohr magneton. If L is particularly large then the distribution of the energy levels in Eq. (1.6) or Eq. (1.7) will be almost continuous and the corresponding sum could be substituted by the integral given the definition of a suitable level density $D(\epsilon)$. In other words

$$\sum_\sigma \sum_{n_1} \sum_{n_2} \sum_{n_3} \rightarrow \sum_\sigma \int d\epsilon D(\epsilon)$$

(1.7)

or, as a free electronic system

$$D(\epsilon) = \frac{2\pi L^3}{\hbar^3} (2m)^{3/2}\epsilon^{1/2} \tag{1.8}$$

The fact that where volume is of the order of L^3 then the level density does not depend on the boundary conditions or shape is known as the Weyl–Laue theorem. This theorem gradually begins to hold good as the volume and quantum number get larger. For particulate metals with a finite volume and a finite number of electrons it is possible to add after volume terms, for example, which are directly proportional to surface area $S \propto L^2$. As a surface effect this must clearly be considered as, for example, surface energy but we shall not take special account of this here.

Questions of bulk are normally dealt with on the basis of Eqs. (1.7) and (1.8). The electrons in each level are distributed in accordance with the Fermi distribution function

$$f(\epsilon) = \left\{ \exp\!\left(\epsilon - \frac{\zeta}{kT}\right) + 1 \right\}^{-1} \tag{1.9}$$

Other values of the electronic system, such as total energy, specific heat and spin field magnetism can now be calculated with ease. However, for reasons given below, these calculations cannot be used in the case of superfine particulate metals.

The principal physical reason to which we must first turn our attention is the electrical neutrality of each individual particle. If we assume that a particulate metal with a radius a has one electron too many or too few then its static electricity gain will be given by the formula

$$W = \frac{e^2}{2a} \tag{1.10}$$

A hydrogen atom with a radius a of 0.053 nm would have a W of 13.6 eV. With a radius of 5.3 nm it would have a W of 0.13 eV and with a radius of 53 nm it would have a W of 0.013 eV. If $W \gg kT$ by comparison with the thermal energy kT of temperature T then this kind of excess or shortfall is unlikely to occur. In fact, even at a constant temperature of 300 K, kT will be no more than 0.025 eV so that a particulate metal in the region of 10 nm will rarely find itself ± 1 electron despite the fact that the total number of conduction electrons could be anywhere from several tens of thousands to several hundred thousand. This fact is evidenced by the electrical conductivity characteristics of thin metallic films.

The second reason is that where the size of a particle is finite then the energy levels given by Eq. (1.5) vary considerably and the energy differences Δ between one energy level and the next are also clearly finite. At absolute zero in the highest energy levels (Fermi levels ζ_0) occupied by electrons, the size of this difference Δ can easily be inferred.

The differences between individual particles in terms of, for example, shape and internal lattice defects are also considerable. The distribution of the energy levels cannot be described with certainty and so we must approach the problem in terms of statistical averages. If we assume the size of a particle to be several nanometers then the number of electrons contained within the particle will be in the order of 10^4 or more and since the quantum number will also be quite high then Eq. (1.8) will give an accurate mean level density. Thus the mean level gap in the region of ζ_0 is

$$\delta \equiv \overline{\Delta} = D(\zeta_0)^{-1} \tag{1.11}$$

With a free electron model it becomes

$$\delta = \frac{4}{3} \cdot \frac{\zeta_0}{N} \tag{1.12}$$

In other words, it can be seen as somewhere in the order of $\delta \sim \zeta_0/N$. Since ζ_0 is in the region of several eV, if we assume $N \sim 10^4$ then $\delta \sim 10^{-4}$ and if we convert this to a temperature value it will be in the region of 1 K.

When we attempt to describe some experimental facts in relation to particulate metal clusters, however, the magic power of the Fermi distribution of Eq. (1.9) is lost. The Fermi distribution is not a law which holds precisely for a given number of electrons (N) but rather holds true only if allowance is made for the statistical fluctuation of N, which is normally determined by ζ. The conditions required for this are not simply that N be large but also that there be a large number of levels in the kT band around the Fermi level. If we turn this round and make

$$\delta > kT \tag{1.13}$$

then the fluctuation between the fixed condition N and the N of the Fermi distribution will become decisively contradictory. Where $\delta \ll kT$ at high temperatures this contradiction has no bearing on the study of heat.

Thus, in particles where electrical neutrality conditions are effective and at low temperatures where dispersion condition Eq. (1.13) holds good, the thermal properties of particle clusters may be expected to differ from those of their corresponding bulk solids. Kubo was the first to point this out in his treatise and for this reason the effect has since become known as the Kubo effect.

An evaluation of the sort of temperatures at which the Kubo effect starts to manifest itself indicates something in the region of 1 K for a particle with a diameter of several nanometers. In other words, at anything less than the temperature of liquid helium the Kubo effect is not apparent.

The sorts of effects that we might expect to see as the intervals between the energy levels get larger are many and various if we think of electrons as existing in the energy levels in pairs [8]. In fine particles with electronic states as shown in Fig. 1.2 the number of electrons in an energy level in a state of excitation varies depending on whether the total number of electrons is odd or even. Where the total number is even then the number of electrons in an energy level is two and where it is odd then the number is one with excess spin also manifesting itself in the latter case. Referring back to the original treatise, we can perform a detailed calculation to find the mean magnetization generated by spin in a weak magnetic field H. For an even number of electrons this would be

$$M \sim 4\mu_B^2 H \cdot \frac{\exp(-\Delta/kT)}{kT} \tag{1.14}$$

Here μ_B is a Bohr magneton and $\Delta = \epsilon_1 - \epsilon_0$. For an odd number of electrons, on the other hand, it would be

$$M \simeq \frac{\mu_B^2 H}{kT} \tag{1.15}$$

The above discussion illustrates the enormous difference between superfine particle band magnetism at low temperatures as in Eq. (1.13) and in the bulk condition. However, the magnetic moment of this type of spin is extremely small, which makes it particularly difficult to measure.

Apart from the Kubo effect which takes account of the existence of the discrete energy levels outlined above, there are numerous other phenomena associated with the concept of superfine particles. For

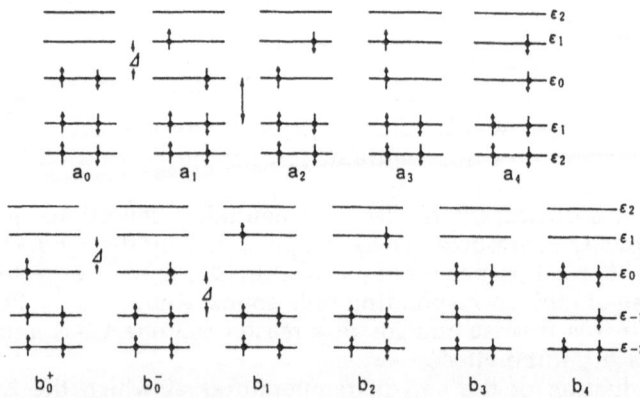

Fig. 1.2 Electronic state of fine particles. a_0 shows the lowest energy state of an even electronic system and b_0^+ and b_0^- show the lowest state for an odd electronic system. a_1, a_2, a_3, a_4 and b_1, b_2, b_3, b_4 illustrate the electronic distribution in an excitation state of even and odd electronic systems, respectively.

example, there is the single magnetic domain particle formed when a highly magnetic particle is reduced in size, the super-paramagnetism which occurs when the particle is reduced even further in size, and a whole range of phenomena which occur when the particle diameter is reduced below that of the mean free movement of an electron or the life span of a superconductive electronic pair.

1.4 Surface Effect in Superfine Particles

The atoms inside a substance are always in a state of balance, being either attracted or repelled by the other atoms around them. The surface atoms, on the other hand, are in a state whereby they are attracted unilaterally towards the interior of the substance by the inside atoms. This means that the surface atoms are in a higher energy state than the interior atoms. This surplus energy is described in terms of amount per unit of surface area (J/m^2, erg/cm^2) and is known as surface energy.

As explained above, superfine particles are said to differ from corresponding bulk substances in that their surface areas are relatively large. To be more precise, the ratio of surface atoms to the total number of atoms in a superfine particle is much greater than that of the corresponding bulk solid. Table 1.2 shows the number of surface atoms in single particles of different sizes as percentages of their totals [9]. It is clear from this table that the smaller the diameter of the particle, the greater the proportion of surface atoms. For this reason the surface energy of a copper particle, for example, such as that illustrated in Table 1.3 is extremely high. The existence of this energy manifests itself in the form of the surface tension of the particle. Since this surface energy is surplus energy it acts in the case of a liquid, for example, to make the surface area as small as possible, thus effectively producing the final, overall shape of the liquid. As with the film of a soap bubble any attempt to expand the surface area is counteracted by the surface tension effect (N/m, dyn/cm), which seeks to reduce the surface area. The reason why a drop of dew on a leaf forms a globular shape is because this represents the smallest possible surface area for the particular volume of liquid in

Table 1.2 Ratio of surface atoms to total atoms in a single particle

Particle diameter (nm)	Total atom count	Surface atoms (%)
10	30 000	20
5	4 000	40
2	250	80
1	30	99

Table 1 3. Diameter and surface energy of superfine copper particles

Edge length	Number of particles per g of atoms	Number of atoms per particle	Weight of a single particle (g)	Total surface area (cm^2)	Surface energy (erg)	Ratio of surface energy to volume energy (%)
5 nm	5.69×10^{19}	1.06×10^{4}	1.12×10^{-18}	8.54×10^{7}	1.88×10^{11}	5.51
10 nm	7.12×10^{18}	8.46×10^{4}	8.93×10^{-18}	4.27×10^{7}	9.40×10^{10}	2.75
100 nm	7.12×10^{15}	8.46×10^{7}	8.93×10^{-15}	4.27×10^{6}	9.40×10^{9}	0.275
1 μm	7.12×10^{12}	8.46×10^{10}	8.93×10^{-12}	4.27×10^{5}	9.40×10^{8}	0.0275
10 μm	7.12×10^{9}	8.46×10^{13}	8.93×10^{-9}	4.27×10^{4}	9.40×10^{7}	0.00275
100 μm	7.12×10^{6}	8.46×10^{16}	8.93×10^{-6}	4.27×10^{3}	9.40×10^{6}	0.000275

question. A large dewdrop forms more of a flattened globe or disk shape. This is an attempt to counteract the increasing force of gravity in the larger volume of liquid and this squeezing of the globe's height is more energy efficient than a reduction of its surface energy as a perfect sphere.

Since solids, however, are collections of atoms with the anisotropic characteristics of crystals the surface energy varies depending on the grain face. Moreover, since the atoms in a solid cannot move around freely in the same way that they can in a liquid it is not possible for them to reduce surface energy in any meaningful way since they cannot alter their shape. It is thus the case that otherwise similar solids may exhibit differing levels of surface energy depending solely on their shape. The shape which reduces surface energy to its lowest possible value is known as the Wulff polyhedron [10]. This polyhedron is selected for its ability to reduce the total surface energy of all surfaces (σ_i) and the total surface area (A_i), in other words $\Sigma_i A_i \sigma_i$, to its lowest possible value. In this polyhedron the ratio between the length of any perpendicular line (h_i) drawn from the center point (Wulff point) to any surface and the energy of that surface is constant such that the relationship h_i/σ_i = Constant holds good. A polyhedron obtained in this way is known as a Wulff polyhedron. This is the ideal shape of a crystalline particle in a state of thermal equilibrium. Figure 1.3 illustates a Wulff polyhedron (section). The geometric operation is, of course, a three-dimensional one and only in special cases will a two-dimensional figure constitute an exact section of a three-dimensional Wulff polyhedron.

Herring [11] uses the Wulff polyhedron to help categorize the different types of equilibrium shapes of superfine particles. The range of possible shapes is shown in Fig. 1.4. A pure liquid form (for example, a small drop of water or oil) is, of course, globular in shape. Liquid crystals adopt the shapes shown in Fig. 1.4a and b. Granular particles differ in shape depending on temperature, with Fig. 1.4e representing the shape adopted at $T = 0$ K and Fig. 1.4c and d the shapes adopted where $T > 0$ K.

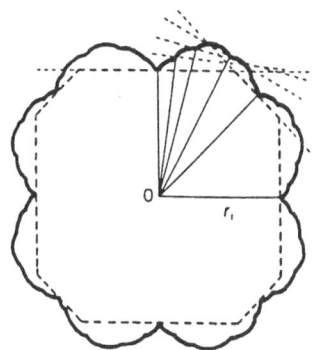

Fig. 1.3 Section of a Wulff polyhedron.

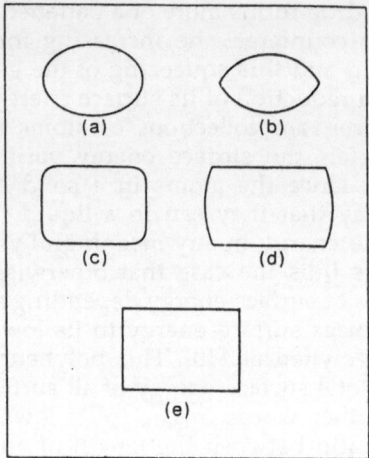

Fig. 1.4 Various shapes which may be adopted by superfine particles.

In an actual solid the movement of atoms is severely limited and they cannot realign themselves in order to reduce surface energy. However, when a particle becomes small enough to be classed as a superfine particle then its total surface area also becomes extremely large and its surface energy becomes high. As shown in Table 1.3, when the diameter of a copper particle reaches a level of 10 nm, its surface energy rises to 2.75% of its volume energy, which is far too great a proportion to be ignored [12]. The limit diameter value used in previous bulk theory was d_c and if we take a ratio of surface energy to total volume energy of 1% as our yardstick then in the case of a particulate metal d_c must be in the region of 30 nm. However, in the calculations in Table 1.3 the agglutinating energy in a gram atom of copper (63.55 g) is taken at 81.5 kcal/mol and the equilibrium surface energy of copper at 2200 erg/cm².

Table 1.3 relates to particles of copper. If we now measure the surface energy of a series of superfine cubic particles of tin(IV) oxide with diameters ranging from 1 μm down to 2 nm then we will obtain a set of results as shown in Table 1.4. The table clearly shows that a particle of 10 nm diameter has 100 times the surface energy of a particle of 1 μm diameter, and also, at roughly 9×10^5 cm²/g, around 100 times the proportionate surface area [13].

As the ratio of surface energy to total particle energy gets larger so even solids can change their external form in order to reduce surface energy. This is an example of the surface effect. The lower melting point of superfine particles or their unusual crystalline structure may be attributable to this surface effect.

The melting point of Au, for example, is clearly dependent on particle diameter, as shown in Fig. 1.5. This figure illustrates the dramatic

Table 1.4 Dependence on particle diameter of surface energy and proportionate surface area of superfine tin(IV) oxide particles

Particle diameter (nm)	Surface energy E_s (erg/mol)	E_s/Total volume energy R	Proportionate surface area (cm²/g)
2	2.04×10^{12}	35.3	4.52×10^6
5	8.16×10^{11}	14.1	1.81×10^6
10	4.08×10^{11}	7.6	9.03×10^5
100	4.08×10^{10}	0.8	9.03×10^4
1 000	4.08×10^9	0.1	9.03×10^3

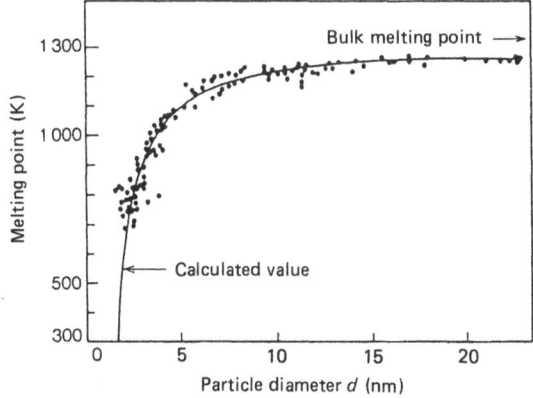

Fig. 1.5 Dependence of melting point of fine particles of Au on particle diameter.

reduction in the melting point of Au as particle diameters drop below 5 nm. For example, a particle with a diameter of 2 nm has a melting point of 600 K. A number of models have been proposed to explain this phenomenon. For example:

1. A system in equilibrium where fine solid and fine liquid particles with identical mass coexist in saturated steam.
2. A system in equilibrium where fine solid particles already exist covered by a layer of fine liquid particles.

These models both correspond extremely closely to actual experimental results [14]. In the second model, however, the thickness of the liquid layer covering the surface of the fine solid particles must be constant without regard for the fine particles themselves.

In that boundary area we refer to as the surface of a particle, not only the configuration of the atoms but also their electronic and magnetic structures are very different from those dealt with by normal solid state theory. This is a point that should not be ignored when considering the characteristics of superfine particles with their proportionately large surface areas, but understanding will not come easily since similar problems have still to be solved in respect of the surfaces of large solids. While superfine particles and large solids may have similar surfaces there nevertheless remain a large number of differences between the two, and one particularly interesting area for future research lies in the fact that superfine particles may have extremely large proportionate surface areas such as are never found in solids.

1.5 Interaction Between Superfine Particles

Up to now we have talked about the surface and the interior of a particle as if they were entirely separate entities, but in reality it may not be appropriate to make quite such a clear distinction. Things which affect the surface atoms of a particle will also have a strong effect on the atoms further inside the particle, with the result that there is even a possibility that the characteristics of the particle as a whole may undergo some change. In groups of particles, in other words, powders, the surfaces of neighboring particles come into direct contact with each other and just as the influence of surface atoms extends to the whole of the particle to which they belong so too does it extend to the whole of the particle to which they are adjacent.

Furthermore, we have so far considered only the ideal situation where a particle is constituted entirely of identical atoms but the reality of the matter is, of course, that superfine particle surfaces are characterized by the presence of a variety of different molecules or atoms. These various molecules, acting either directly or indirectly through the interface between particles, must undoubtedly interact with each other.

The direct effect involves perhaps the adhesion or bonding of heterogeneous molecules to the surface of a particle. This may not involve more than a single layer of atoms at the surface but, as was explained above, in the case of a superfine particle this single layer may account for a substantial proportion of the total number of atoms which constitute the particle. This is certainly enough to have a profound influence on a powder's observable characteristics. Depending on the surface form of the particle in question there are cases where atoms which have adhered to the surface may even make up the majority of the atoms in the particle.

Indirect influences may include such effects as abnormal optical properties or inductive capacity resulting from interaction between particle surface and medium. As a phenomenon which applies specifically to surfaces it may be expected to be particularly noticeable in the case of superfine particles with large proportionate surface areas.

The extremely short distance between the surface and the center of a superfine particle permits an atomic medium or particle close to the surface of another particle to exert an influence right through to the center of that particle. It is eminently possible for the whole of a particle to fall within the ambit of, for example, van der Waals' forces, a strong magnetic field or an ionized electrical field. Superfine particulate metals commonly agglomerate with each other and once this has taken place it is not at all easy to disperse them again. This is also thought to be due to some form of this proximity effect.

In agglomerations of superfine particles, the particles themselves sometimes bond tightly with each other in such a way that the powder as a whole becomes glutinous, making it difficult to handle.

Infiltration into the interior of a particle is not limited solely to the action of a force but also probably involves the active transfer of atoms between the interior and exterior of a superfine particle through the interface represented by its surface. It is eminently possible that alien atoms may infiltrate a crystalline lattice and thereby necessarily alter the overall structure of the crystal.

The interaction of superfine particles is not confined to adjacent particles but may also occur where superfine particles are dispersed within a medium. Nor is the effect limited to that of the medium on the surface atoms of a large, solid particle but may equally extend from the medium through the whole of the particle. The types of interaction that may occur is an extremely interesting area of solid state physics but one which has as yet received only scant attention.

We will now cite three further striking examples of interaction between adjacent particles or particle surfaces. The first example relates to the formation of a tunnel effect or Josephson junction between particles in a film produced by the formation of particles of a superconductive metal by vacuum evaporation [15]. Within the vacuum the metal evaporates onto a dielectric substrate and initially forms into what is called the island structure, of entirely discrete superfine particles. Given the appropriate evaporation conditions a tunnel current will flow between the metal islands enabling the formation of countless metal–dielectric–metal (M–I–M) diodes. This type of island-structured, thin metallic film diode can be used to make a whole variety of high speed wave detectors and frequency mixers, handling anything from microwaves through to mm-waves and on into the infrared zone. Wave detection characteristics are optimized where the metal islands are formed from metals with comparatively high melting points such as silver or silver palladium. The size of the island particles can be anything up to 10 nm with around 1–2 nm between islands.

The second example relates to a particulate tin film produced by evaporation onto a fused quartz substrate in a comparatively weak vacuum of approximately 10^{-3} Pa. Countless Josephson junctions form between the particles of the film, connecting the particles at random both in series and in parallel, thus facilitating their use as highly sensitive wave detectors and mixers capable of handling anything from microwaves through to waves in the remote infrared zone. When this type of thin film element is used in the 10GHz waveband the resulting voltage performance index is $R_v \simeq 9 \times 10^{16}$ [V/W], [NEP]$\simeq 3 \times 10^{-13}$ [W/$\sqrt{Hz}$] [16].

For the final example we will offer a brief outline of supranormal magnetism resulting from magnetic interaction [17]. It is well known that highly magnetic substances such as Fe, Co and Ni, along with their compounds, adopt a multi-magnetic domain structure in their normal bulk state. Each domain is separated from the rest by domain walls and the spin which supports the magnetic strength slowly changes their direction within. Reversal of magnetization occurs as a result of the movement of these domain walls. It is estimated that the thickness of the walls is generally in the region of 100 nm. A superfine particle may well have a diameter of less than the thickness of these domain walls and in such a case it will assume a single domain structure where the reversal of magnetization will result from a complete spin induced by the movement of the domain walls. Changes in magnetic coercive force and susceptibility will thus occur as shown in Fig. 1.6 [18].

Carefully selected Fe, Co and Ni compounds can have a magnetic coercive force in the order of $H_c = 2500$ Oe. In this case particle size is 20 nm. With 30 nm particles of Fe or Ni the maximum possible magnetic coercive force is 1700 Oe. If particle size is reduced further then the state

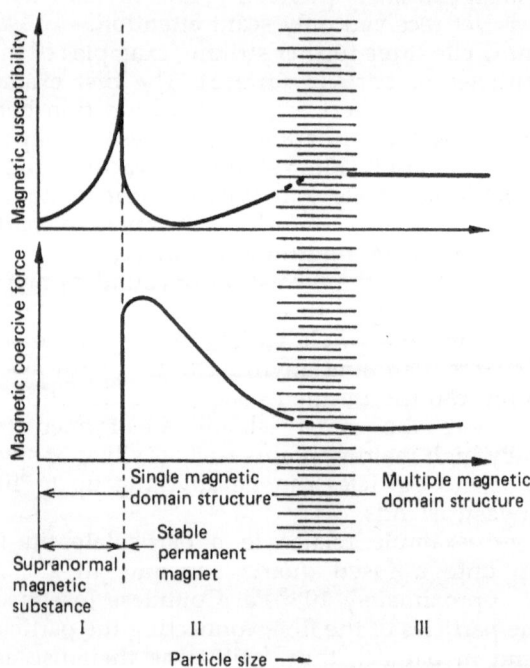

Fig. 1.6 Dependence on particle size of magnetic coercive force and magnetic susceptibility in highly magnetic substances.

of equilibrium which exists between a particle's anisotropic energy K_u and its thermal energy is upset, and the spin is affected by thermal fluctuation and begins to exhibit supranormal magnetic behavior. In other words, the strength of the thermal influence grows beyond that of the mutual magnetic interaction between particles.

As explained above, therefore, in the case of superfine particles surface or interface phenomena will, generally speaking, have a proportionately greater significance than is the case with larger particles. However, since our observations of superfine particles relate in reality to agglomerations or "powders" it must thus be clearly understood that there is also likely to be an element of interaction between particles incorporated in our findings.

References

1. Shin Gijutsu Kaihatsu Jigyoudan: Summary of address on Hayashi Cho-biryushi Project Research Report (in Japanese)
2. N. Wada: Ceramics, 19, p. 464 (1986) (in Japanese)
3. Nihon Funtai Kogyou Gijutsu Kyokai (ed.): Cho-biryushi Ouyou Gijutsu, p. 2, Nikkan Kogyo Shinbunsha (1986) (in Japanese)
4. K. Kimura: Zairyou Gijutsu, 5, p. 350 (1987) (in Japanese)
5. Nihon Funtai Kogyou Gijutsu Kyokai (ed.): Cho-biryushi Ouyou Gijutsu, p. 4, Nikkan Kogyo Shinbunsha (1986) (in Japanese)
6. Chemical Society of Japan (ed.): Cho-biryushi Gijutsu, p. 1, Gakkai Shuppan Center (1985) (in Japanese)
7. R. Kubo: J. Phys. Soc. Japan, 17, p. 975 (1962)
8. R. Kubo, Kawahata: Nihon Butsurigakkaishi, 23, p. 718 (1968) (in Japnaese). R. Kubo: Kotai Butsuri Bessatsu Tokushugou, p. 4 (1984) (in Japanese)
9. R. Ueda: Kotai Butsuri Bessatsu Tokushugou, p. 1 (1984) (in Japanese)
10. K. Kimoto: Kotai Butsuri Bessatsu Tokushugou, p. 68 (1984) (in Japanese)
11. C. Herring: Structure and Properties of Solid Surface (ed. R. Gomer and C.S. Smith), University of Chicago Press (1953)
12. K. Kimoto: Ryushi Kouzou (K. Hayakawa (ed.): Funtai Bussei Sokuteihou), p. 1, Asakura Shoten (1973) (in Japanese)
13. Ogawa, Nishikawa, Abe: CPM 81–28, p. 37, Denshi Tsushin Gakkai (1981) (in Japanese)
14. P. Buffat, J.P. Borel: Phys. Rev. A13, p. 2287 (1976)
15. Tamiya, Okabe: Kotai Butsuri Bessatsu Tokushogou, p. 138 (1984) (in Japanese)
16. N. Fujimaki, Y. Okabe, S. Okamura: J. Appl. Phys., 52, p. 912 (1981)
17. M. Oda: Kotai Butsuri Bessatsu Tokushugou, p. 103 (1984) (in Japanese)
18. S. Iida: Koushitsu Jisei Zairyou, p. 15, Maruzen (1976) (in Japanese)

Additional References

19. T.A. Ring, "Processing of fine ceramic powders", MRS Bulletin, pp. 34–40 (Jan. 1990)
20. E. Matijevic, "Fine particles: science and technology", MRS Bulletin, pp. 18–20 (Dec. 1989)

Chapter 2

Physics of Superfine Particles

Japan leads the world in research on the physics of superfine particles and has made many scientific and technical discoveries in this field. In recent years, there has been a growing demand, particularly in the industrial world, for the development of new materials. This demand has focused on superfine particles, spurring on the practical application of current knowledge about their properties.

Superfine particles possess many extremely interesting properties not found in conventional materials. By combining the manufacturing and handling methods described in Chap. 4, it should prove possible to use these particles for practical purposes.

The research on superfine particles carried out by the Exploratory Research for Advanced Technology was aimed at exploring their properties from new angles, and finding ideas for new applications.

2.1 Introduction

There now seems to be a general consensus that the term "superfine particles" designates particles smaller than 0.1 μm (100 nm). For the purposes of this research, the lower size limit of the superfine particle was set at the upper size limit of the cluster:[1] a particle diameter of 1–2 nm (comprising between several tens and several hundreds of atoms). However, from a practical standpoint, it may be difficult to adhere strictly to this definition.

Many physics researchers and other scientists first became inspired to study superfine particles after reading about observations made by a

---

[1] The cluster, an ultra-superfine particle comprising a countable number of atoms, is being studied by physicists and chemists.

physics research group, led by Ryoji Uyeda (Professor Emeritus at Nagoya University) using an electron microscope, and after seeing the electron micrographs later published by Professor Uyeda and Kazuo Kimoto (Professor of Liberal Arts at Aichi University). There has recently been an upsurge of scientific interest in these particles among industrial scientists and scientists in many other fields.

Table 2.1 shows the relationship between the diameter of a superfine particle and the number of atoms it comprises, and the ratio of surface atoms to the total number of atoms. The atoms on the inside of a superfine particle are in contact with other identical atoms, but the atoms which form the surface of the particle come into contact, on one side, with gases, metals and other substances. When the particle diameter is 5 nm, about 40% of all the atoms in the particle are on the surface. This means that if particles of this size are to be used for practical purposes, a thorough knowledge of the properties of the particle surface and related phenomena, and expertise in handling such particles, will be vital.

Particles of this size, however, are the most likely to demonstrate (what we hope will be) the characteristic properties of superfine particles, so in the future superfine particles may be divided into two groups (for example 10 to 100 nm and 1 to 10 nm), and their physical properties studied separately.

Superfine particles, which are intermediate between the bulk (solid) state and the atomic state, differ from bulk state materials in their volume (size) and surface effects.

Volume effects occur in the following situations: (a) when the particle is smaller than the magnetic domain of a bulk material magnetic substance; (b) when the particle is smaller than the mean free path of the electrons in a conductor; and (c) when the particle is smaller than the wavelength of light. These volume effects cause superfine particles to behave differently to conventional materials, and are thus problematic.

There are also surface effects: the specific surface area[2] of the (entire) particle is inversely proportional to the particle diameter, so in the superfine particle range, the specific surface area becomes very large –

Table 2.1 Ratio of surface atoms to total number of atoms in a superfine particle

Diameter (nm)	Number of atoms	$\dfrac{\text{Surface atoms}}{\text{Total no. atoms}}$ (%)
20	2.5×10^5	10
10	3×10^4	20
5	4×10^3	40
2	250	80
1	30	99

[2] Surface area per unit weight of particles; normally expressed as surface area per gram.

somewhere in the range from a few to several hundred square meters per gram. This value is 10 times greater than particles now in practical use (which range in size from several tens to several hundreds of micrometers. As is evident from the electron micrograph taken by Sumio Iijima (Research of the Hayashi Superfine Particle Project of the Research Development Corporation of Japan, currently with NEC Corp.) (Fig. 2.1), the surface of a superfine particle is stepped (showing the layers of atoms), and this is the main reason why the surface properties of superfine particles have such a strong effect during catalysis and sintering [1].

In addition to observations like this, carried out using an electron microscope (transmission type), many recent studies on the applications of separate superfine particles have been concerned with the forms and arrangements assumed by groups of superfine particles.

The properties of superfine particles and possible applications are shown in Table 2.2. This chapter focuses mainly on these properties.

2.2 Structure and Form

The work of Uyeda, Kimoto et al., centering on the use of a transmitting electron microscope, has played a leading role in basic research on the properties of metallic superfine particles. Based on their observations from electron micrographs of metallic superfine particles, this team reported [2] on the marvellous crystal habit[3] which was found to occur when the argon atmosphere at the time of formation contained no oxygen

Fig. 2.1 The surface of a superfine particle of Al_2O_3 (alumina) as seen through a powerful electron microscope [1]. The black points are aluminum atoms, lined up neatly inside the superfine particle, and forming the stepped surface. (Photograph courtesy of Nobuo Iijima.)

---

[3] Crystal habit: the directional characteristics of crystals.

Table 2.2 Properties of superfine particles and applications planned for the future

Property	Application
Particle size	
Single magnetic domain	Magnetic recording
Mean free path of electrons in a solid	Special conductors
Smaller than light wavelength	Light or heat absorption
Formation of ultrafine pores due to aggregation of superfine particles	Molecular filters
Uniform mixture of different kinds of superfine particles	Research and development of new materials
Surface structure at the atomic level	Chemical catalysts
Surface area	
Large specific surface area	Catalysts, heat-exchange materials
Large surface area, small heat capacity	Combustion catalysts, sensors
Lower sintering temperature (lower fusion point)	Sintering accelerators

(i.e. when the argon gas was highly pure) and when the particle size was between 20 and 30 nm.

Superfine particles are generally thought to be spherical, but electron micrographs show that superfine particles of magnesium, for example, take the form of hexagonal discs, and selected area diffraction has demonstrated that these hexagonal plates are single crystals of metallic magnesium [3]. Similarly, interesting electron micrographs have been produced for metals such as manganese, chromium, iron, silver and beryllium [4–6].

A recent development in the observation of superfine particles using the electron microscope is the use of new, high-resolution, ultra-high vacuum electron microscopes. This technique has been used by Iijima in his work on the Exploratory Research for Advanced Technology, and has been highly acclaimed for the opportunity it gives scientists to observe superfine particles from a new angle (it is particularly useful in basic research on catalysts and sintering, for example) [7].

Iijima's new observations, based on enlarged electron micrographs (magnified approximately 5 million times on photographic paper so that 1 nm in actual size becomes 5 mm in the photograph) include the following: (a) the surface of superfine particles which at first glance appear to be spherical is in fact stepped (due to layers of atoms); (b) atoms are arranged in an orderly fashion inside each particle; and (c) atoms move on the surface of superfine particles. Even more interest was attracted by an experiment in which a TV camera was connected to the electron microscope mentioned above, as in Fig. 2.2, and a metallic superfine particle (consisting of about 450 atoms and with a diameter of

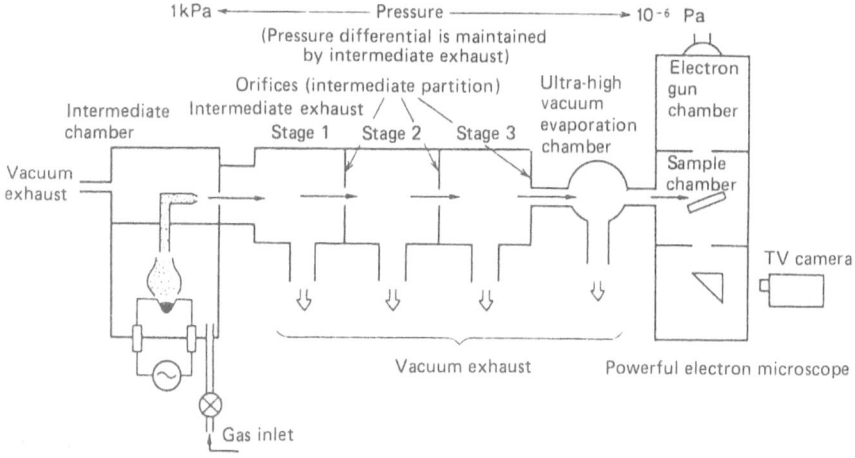

Fig. 2.2 Schematic diagram of sample preparation chamber connected directly to electron microscope [7].

about 2 nm) in the electron microscope sample chamber was projected onto a TV screen – for the first time in world history – and was seen to change its shape and structure from moment to moment, almost like a living cell [8].

These images widened the circle of researchers interested in superfine particles. Iijima's work is explained and the electron micrographs are described in more detail in [9] and [10].

A new experimental technique for observing superfine particles using an electron microscope was devised by the Uyeda team. Newly formed superfine particles (which have not come into contact with the atmosphere) are introduced into the sample chamber of an electron microscope by passing a beam of superfine particles through a vacuum: this makes it possible to observe superfine particles which have not come in contact with air. The interest of this technique also lies in the fact that it allows superfine particles to be observed a very short time after they have been formed. Figure 2.2 shows a schematic diagram of the superfine particle beam generator which is connected to the electron microscope.

The term "fine particle beam" was first coined for the beam designed by Uyeda, which shoots out superfine particles through a vacuum in a narrow parallel beam. It refers to beams of superfine particles with a diameter larger than 5 nm, as opposed to the cluster beams which are now the subject of active research in the West [11, 12].

Generators like the one shown in Fig. 2.3 are based on the principle of this fine particle beam. In this device, superfine particles are created by evaporation in gas inside an evaporation chamber, and this mist – superfine particles mixed with the gas – is drawn into an intermediate

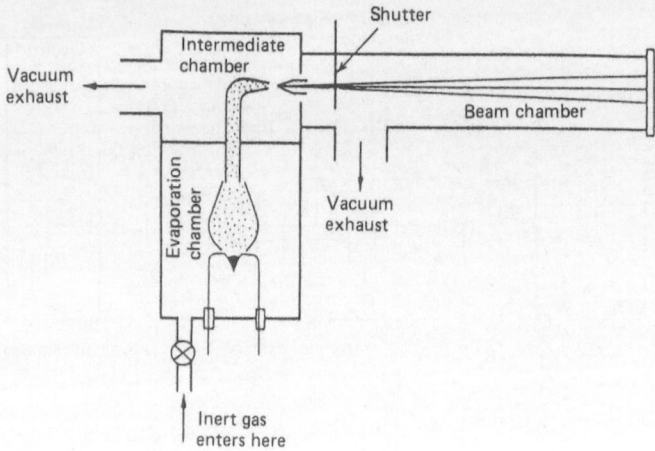

Fig. 2.3 Superfine particle beam generator [11].

chamber and exits through a small hole. The gas is then removed using a vacuum pump, whereupon the particles shoot forward.

The superfine particles, which have only just been formed by evaporation, thus arrive in the electron microscope sample chamber shown in Fig. 2.2, where they can be observed in a new way.

Chikara Hayashi (Chairman of ULVAC JAPAN, Ltd, overall head of the Hayashi Superfine Particle Project[4] and Chief Supervisor of the Exploratory Research for Advanced Technology) claims that these new methods of observing superfine particles using the electron microscope go beyond mere observation of the form and structure of the particles themselves and offer definite advantages as new means of observing material surface behavior, to be used in research on material surface reactions [13].

2.3 Properties

Strictly speaking, the physical properties of (individual) superfine particles should really be determined by scientific study of the characteristics possessed (or shown) by single superfine particles. In reality, as we stated

[4] Over a five-year period from October 1981 to September 1986, research on superfine particles was carried out by four research groups, which studied basic properties, physical applications, biological and chemical applications, and production methods, respectively.

earlier, studies aimed at developing practical applications have dealt with aggregates of superfine particles (which should be termed superfine powders).

2.3.1 Magnetism

The use of magnetic bodies of superfine particle size, to provide magnetic recording materials with improved magnetic properties, was one of the earliest applications of superfine particles.

In magnetic particles larger than superfine particles, the shifting of the boundaries between magnetic domains (areas in which the direction of the magnetism is consistent) within each particle cause magnetic reversals to occur constantly in the weak magnetic field. Superfine particles with a diameter of the order of several tens of nanometers have one magnetic domain per particle, that is to say, the whole of the particle is a permanent magnet, magnetized in one direction: to reverse its magnetism requires a complete reversal of the magnetism of the whole particle, not just a reshuffle of magnetic domains. If they are able to exert a strong repulsive force, particles of this type are termed single magnetic domain particles.

Magnetic recording density is determined by the number of magnetic particles inside one recording unit [14]. By lining up magnetic domain particles of this type in the direction of magnetism, the improvement in magnetic properties can be investigated.

Akira Tasaki (Professor of Physical Engineering at Tsukuba University) et al. experimented with evaporation and recovery within a magnetic field in order to produce magnetic superfine particles which fulfilled the above conditions. The Tasaki team used an experimental resistance-heating gas evaporation device (see Sect. 4.3.2 in Chap. 4). The superfine particles generated were chainlike and aligned in the direction of magnetization. As for their magnetic properties, the third quadrant of the hysteresis curve, which shows the $B-H$[5] properties, was greatly enlarged. The reason for the increased angularity[6] was investigated [14, 15].

It was considered that the individual superfine particles are given magnetic anisotropy during the production process, and when they are linked together in chains, the magnetic anisotropy can be used. Superfine particles of magnetic material, which have been passed through a magnetic field directly after evaporation, are linked together in a chain, as shown in the electron micrograph in Fig. 2.4. (Superfine particles were generated by evaporation in a gas phase as discussed in Chap. 4, where this chainlike formation is described.)

---

[5] The curve which shows the relationship between the strength of magnetization (B) and the magnetizing force (H).

[6] $B-H$ properties: the proportion of residual magnetic flux density (B_r) relative to the maximum magnetic flux density.

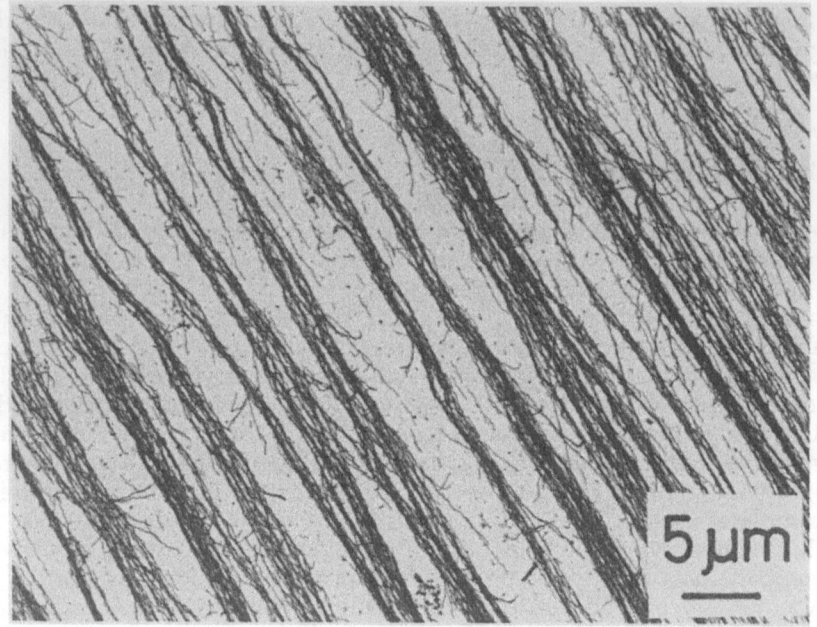

Fig. 2.4 Superfine particles of magnetic material, linked together in chains (enlarged photograph shown in Fig. 4.8, Chap. 4).

A study was carried out in which magnetic recording tape was coated with these magnetic superfine particles, and the tape's performance as audiotape, videotape and computer tape was then investigated. Kaoru Imanishi (Matsushita Electric Industrial Co. Ltd) et al. have reported on the properties of the master tape used for the reproduction of VHS video tape. To make this tape, superfine particles (produced by the evaporation method) of a magnetic alloy with the chemical composition Fe–Co were used. The coercive force was 2060 Oe, and the residual magnetic flux density output was 3250 G. The B–H curve is shown in Fig. 2.5, and shows a high output of about 7 dB[7], higher than conventional cobalt-added master tape. This tape also had a higher output than VHS video slave tape – about 11 dB at 4 MHz. Its characteristics are shown in Fig. 2.6 [16, 17].

Imanishi et al. pointed out that the superfine particles they had used, which were made from magnetic materials, were very uniform in diameter, and assumed a chainlike form to produce a material with excellent magnetism.

---

[7] Decibel: in electric output, for example, 20 times the common logarithm for the ratio for opposing values. For example, 6 dB is equivalent to twice the output.

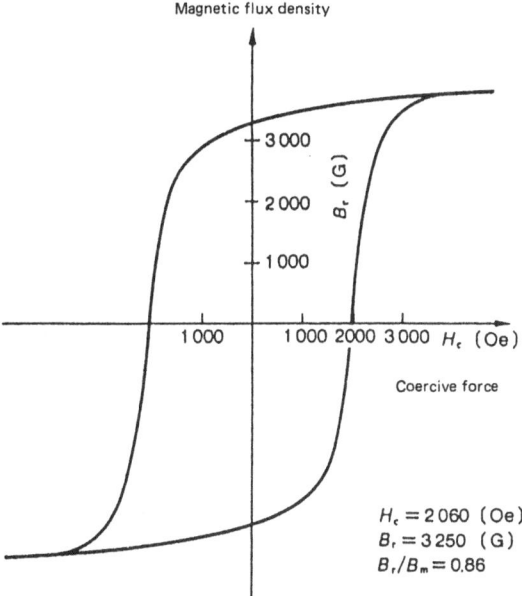

Magnetic flux density

$H_c = 2060$ (Oe)
$B_r = 3250$ (G)
$B_r/B_m = 0.86$

Fig. 2.5 B–H properties [16, 17] for VHS master tape using superfine particles of magnetic alloy (chemical composition: Fe–Co).

2.3.2 Lowering of Melting Point and Sintering Onset Temperature

As is clear from Table 2.3, the energy on the surface of a particle is inversely proportional to its diameter. When the diameter of a superfine particle is less than 10 nm, the surface energy makes up almost 10% of the combined energy of the whole, and the influence of surface energy on the physical properties is no longer negligible. This fact is reflected in phenomena observed in superfine particles, such as the lowering of the melting point, the onset of sintering at a temperature lower than that of conventional particles, and the improved (surface) reactivity.

It has often been noted that the finer the superfine particle, and the smaller the particle diameter, the lower the melting point will be. Mieko Takagi (Tokyo Kasei Gakuin University) has calculated the relationship between the decrease in diameter and the drop in melting point for lead particles. Her conclusion is that a spherical particle of lead with a diameter of 20 nm can be expected to melt at a temperature 15°C lower than bulk state lead (the melting point of lead in its bulk state is 327°C) [18]. Superfine particles have a lower melting point because their volume is much smaller than bulk state lead, so the extra amount of

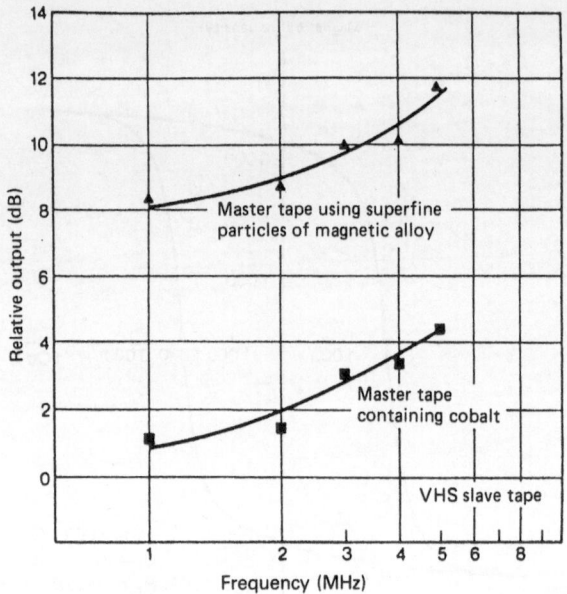

Fig. 2.6 Output over frequency properties [16, 17] for master tape using superfine particles of magnetic alloy, in comparison with VHS slave tape (slave tape: blank videotape).

Table 2.3 Relationship between the diameter of superfine and fine particles of copper, and their surface energy and specific surface area

Diameter (nm)	Specific surface area (m^2/g)	Surface energy (Cal/mol)	$\dfrac{\text{Surface energy}}{\text{Total energy}}$
1	660	1.4×10^4	1.7×10^{-1}
10	66	1.4×10^3	1.7×10^{-2}
100	6.6	1.4×10^2	1.7×10^{-3}
1 000 (1 μm)	0.66	1.4×10^0	1.7×10^{-4}

internal energy due to fusion is very small. This calculation shows that when the (fine particle) diameter is 2 μm, the melting point is 0.15°C lower, and when the (fine particle) diameter is in the order of one micron, there is no drop in melting point.

Figure 2.7 illustrates the relationship, shown by Wronski, between the diameter and melting point of superfine particles of gold. From a particle diameter of 10 nm downwards, a sharp drop in melting point appears [19].

Not only is the melting point lower in superfine particles, but sintering also begins and proceeds at a lower temperature than for conventional particles. This is another aspect of superfine particles which holds promise

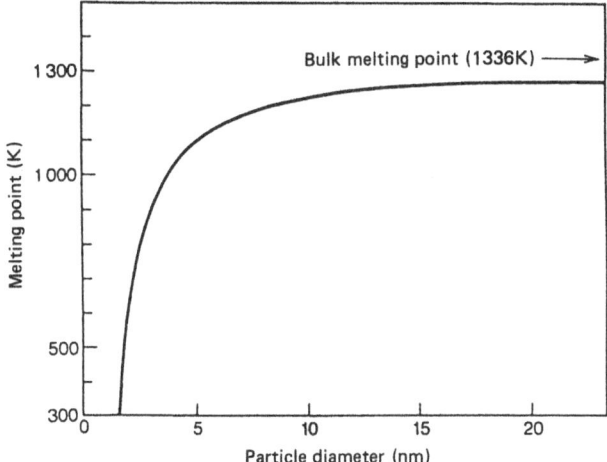

Fig. 2.7 Calculated value for melting point over particle diameter for superfine particles of gold [19].

for practical applications, and was discovered in the early stages of comparative research on superfine particles of metal.

This technique is chiefly used to study post-heating dimensional changes (contraction) in superfine particles exposed to the atmosphere and formed into pellets by molding or using a metal plate of similar methods [20]. This phenomenon is shown in Table 2.4. The sintering onset temperature for superfine particles (all of which have been exposed to the atmosphere) of materials other than gold is the same in all cases – 200°C.

As for the onset of sintering in superfine particles of gold, studies have been carried out in which disk-shaped [21] and filmy [22] particles were generated and changes in their electrical resistance investigated. A decrease in electrical resistance has already been shown at temperatures near to room temperature. One of these studies was carried out by Saburo Iwama (Professor at Daido Kogyo University), who measured the electrical resistance of filmy formed from superfine gold particles which had not come into contact with the atmosphere [22].

Eiji Kamijo (then of Sumitomo Electric Industries, Ltd; currently of Nissin Electric Co., Ltd) set up a high-temperature heating stage in an electron microscope test chamber and used the electron microscope to observe intermittent changes, caused by heating, in the behavior of superfine nickel particles (diameter 20 nm) [23]. Kamijo's report states that, at the low temperature of 200°C, some of the clusters of superfine nickel particles instantaneously grew into large particles, a phenomenon which is different from the contact agglomeration between particles during the sintering of particles and powders, but exactly analogous to the

Table 2.4 Onset of sintering in superfine metal particles

Superfine metal particles	State at onset of sintering
Cu	Pellets of superfine copper particles began to contract at 200°C. The smaller the particle diameter, the greater the contraction [20]
Fe	Superfine iron particles with an average diameter of 50 nm were made into pellets with a diameter of 14 mm and a depth of 2–3 mm and heated in a vacuum. The decrease in their volume was investigated Their volume began to decrease at between 300 and 400°C – over 200°C lower than for bodies formed from fine particles whose size was in the order of microns [21]
Au	The onset of sintering at between 60 and 80°C in superfine gold particles with an average diameter of 20 nm was investigated through variations in electrical resistance [21] Membranes formed from newly-generated superfine gold particles decreased in electrical resistance even below room temperature [22]
Ni	Superfine nickel particles with an average diameter of 20 nm were heated while under observation using an electron microscope. At about 200°C, some of the superfine particles were observed to agglomerate and coalesce [23]

phenomenon which occurs when several drops of mercury coalesce [24].

These phenomena have improved our understanding of the way superfine particles merge together at low temperatures. However, the most interesting and promising aspect of the sintering of superfine particles, and the aspect which we are most curious about, is the behavior shown by superfine particles which have not been exposed to air after being generated (and which therefore have a comparatively clean surface). We hope that methods of handling superfine particles will be established which will shed light on this subject.

All the examples above concern superfine metal particles, but there is also a great deal of interest in the sintering phenomena of superfine particles of chemical compounds. Concerning the production of ceramic particles, for example, plans are afoot to produce sintered bodies with improved characteristics by using finer particles, and the use of superfine particles will probably spread to other materials and new applications.

2.3.3 Optical Properties

It is known that when metals are evaporated in a suitable low-pressure atmosphere (gas evaporation), an evaporated substance (superfine particles) is obtained which is black and absorbs light well. This substance has advantages over the highly reflective shiny surface formed by a metal evaporation surface created within a high vacuum. The black evaporated substance (i.e. superfine metal particles) is known as metal black, and research has been carried out into the optical properties of the metal black which has the best absorbency of radiation, namely Gold Black.

In brief, gold black is made by using a vacuum evaporation device which works by the resistance heating method, and evaporating metallic gold in a W heater in N_2 gas under low pressure. It has been observed that when the N_2 gas pressure is 400 Pa, the size of the superfine gold particles is 25 nm, and at 40 Pa, the particle size is 5 nm [25].

Figure 2.8 shows the spectral reflectivity of gold black, within the visible spectrum, in comparison with other blacks (lamp black and black lacquer). Figure 2.9 shows measurements of the spectral reflectivity of gold black in the visible spectrum [25].

Kazuyoshi Ito (Electrotechnical Laboratory) has compared the properties of gold black with many known properties of metal blacks, and has shown that it is an excellent blackening substance with a large infrared absorption rate/thermal capacity ratio [26].

Ito has stated that the conditions and properties needed to synthesize gold black of this type are now known, and that optical applications have already been established: for example, the use of gold black as a coating for the light-receptor surface in a radiation detection device.

Research on gold black, which has taken place outside the mainstream of research on superfine metal particles, is most interesting in that it links up with superfine gold particles produced by gas evaporation.

2.3.4 Active Surface

It is fully expected that as the particle diameter gets smaller, the proportion of aligned crystal surfaces appearing on the particle surface will decrease,

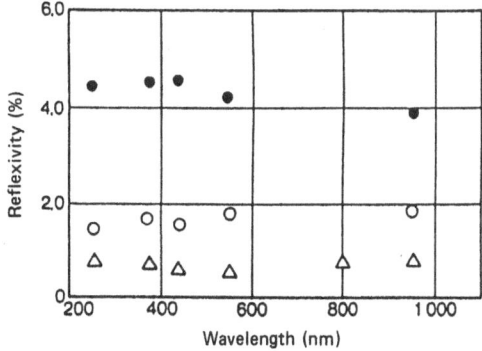

Fig. 2.8 Spectral reflectivity of black absorbent bodies near the visible spectrum. ●, lamp black; ○, black lacquer; △, gold black (superfine gold particles).

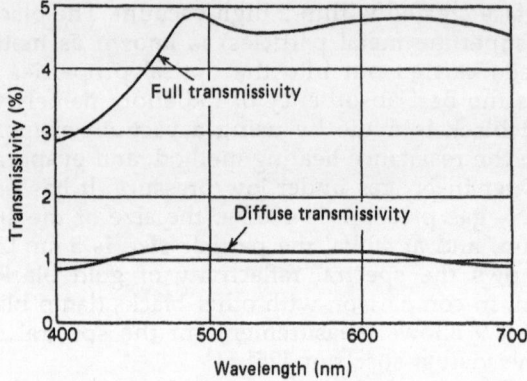

Fig. 2.9 Spectral transmissivity of gold black (superfine gold particles) within the visible spectrum.

and the ratio of surface atoms with unsaturated linkages will increase [27]. In situations involving a catalytic active substance placed on a carrier substance to be used as a solid catalyst, many ways have been found to make the catalyst particles as fine as possible.

For superfine particles of a size under 100 nm (0.1 µm), within the size range which can be observed using an electron microscope (by persons who are not experts in the use of this device) (i.e. down to a particle diameter of 2 nm), the particles dealt with in catalytic chemistry in the past have been towards the smaller end of this range [28].

The process of research and development on superfine particles using gas evaporation – the typical physical production technique – is very different from the chemical colloid production technique which analytical chemists are now working on. There is therefore a need for experts in techniques for handling superfine particles, including support techniques and surface processing techniques.

For example, the surface of superfine nickel particles is slowly oxidized because they are handled in air after being synthesized (slow oxidation is described in Sect. 2.3.5), but to use it as a hydrogenation reaction catalyst, the oxidized surface layer[8] must be removed to leave a particle with a diameter of 30 nm (BET specific surface area: 27 m^2/g).

Experiments have been carried out in which superfine particles produced by the physical method were used as catalysts: superfine particles of tungsten carbide have been used as a hydrogenation catalyst similar to platinum in experiments involving fuel cell electrodes [29], and superfine nickel particles have been widely used as a hydrogenation reaction catalyst. Monoene was obtained from conjugate diene with a

---

[8] Superfine nickel particles which have been subjected to slow oxidation processing have an oxidized surface layer several atoms thick.

higher selectivity than with the conventional catalyst, Raney nickel.[9]

Using gas evaporation, bimetallic superfine particles (i.e. superfine particles with more than one component) can be produced. For example, experiments have been carried out in which Cu–Zn particles produced by this method were used as a catalyst for the synthesis of methanol from CO and H_2 [31]. (See Chap. 4, Sect. 4.3.2.f for a description of the production method.)

In superfine particles produced by gas evaporation, the active catalytic component of a metal such as Ni, Co or Fe showed a spherical and sharp granular distribution, had few surface impurities, and was suitable for practical use as bimetallic particles. Experiments will therefore be carried out to find out how well they work as catalysts, and they show promise as a new material for regulating catalysts.

2.3.5 Surface and Surface Stability

The smaller the particle, the greater the surface area per unit weight (termed specific surface area). The specific surface area of a particle is usually measured by the BET adsorption method. When the individual particles are spherical in shape and the particle diameter is comparatively uniform, the average particle diameter can be calculated from the measurement of the specific surface area.

Superfine particles produced by gas evaporation conform to these conditions. As shown in Table 2.5, the average particle diameter measured from highly enlarged electron micrographs[10] (taken to determine the granularity distribution of the superfine particles and the average particle diameter) is almost the same as the particle diameter calculated from the measurement for the specific surface area, described above.

When superfine particles with a large surface area had been produced in this way, and before they had been exposed to the atmosphere (i.e. before they had come into contact with air), the process known as slow oxidation was carried out, preventing oxidation from progressing too rapidly. The superfine particles produced had an extremely large surface area in proportion to their volume. For this reason, newly-produced superfine particles, which have a "fresh" surface, need to be processed in such a way that the particles, which have an extremely small thermal capacity, do not undergo a rapid increase in temperature due to the heat generated by oxidizing the surface layer. It has been established that once the particles have been subjected to this treatment, oxidation does not progress further when the particles are left at room temperature.

---

[9] A blackish nickel powder catalyst made by the liquid phase method and in wide practical use. Named after its inventor, Raney.

[10] In an electron micrograph where the image is magnified 200 000 times, a particle diameter of 10 nm corresponds to 2 mm in the photograph, so in most cases it is possible to measure the particle diameter.

Table 2.5 Average particle diameter and specific surface area for typical (commercially available) superfine particles

Superfine particles	Cu	Fe	Ni	Co
Average particle diameter indicated* (nm)	50	30	30	30
Specific surface area measured† (m²/g)	12.9	38.5	26.5	21.7
Average particle diameter calculated‡ (nm)	52	20	26	32

* Value measured from enlargement of electron micrograph.
† Value measured by the BET adsorption method.
‡ Value calculated from the specific surface area measured using individual spherical particles (from Shinku Yakin KK catalog).

Tasaki has investigated the progress of oxidation in these superfine particles, studying changes over time in the magnetic properties of magnetic superfine metal particles. In the early days of research into superfine particles (this research began in the 1960s), the measurement of the magnetism of strongly magnetic superfine particles was used as an auxiliary method for measuring the layer of oxide on the surface of the particle [15]. Through these measurements, it was noticed that in superfine metal particles produced by gas evaporation, saturation magnetization did not decrease even over long periods of time [32]. Even when superfine iron particles with a particle diameter of 20 nm were left (exposed to air) in a laboratory for almost a year, measurements showed that oxidation had not advanced any further. This phenomenon is one of the main reasons why superfine magnetic metal particles produced by gas evaporation were deemed to be suitable for use as high-density recording materials. This is reflected in their outstanding durability as magnetic recording materials, and in the progress made in the development of high-denisty magnetic recording tape, to be described later.

To give an example, Kosaburo Nakamura (NTT Telecommunications Laboratories, Materials Technology Laboratory) et al. carried out a comparison of tape durability between magnetic computer tape made with superfine particles of iron–nickel alloy produced by gas evaporation, and magnetic material made with fine iron particles produced by conventional methods. In this comparison, the drop in the magnetic flux density of the tape, due to changes over time, was investigated. As can be seen in Fig. 2.10, superfine particles with a small diameter and a large surface area in relation to volume showed excellent durability [33].

The question as to why oxidation progresses so slowly on the surface of superfine metal particles made by gas evaporation will continue to receive scientific scrutiny in the future; meanwhile, changes over time in superfine metal particles subjected to slow oxidation treatment are, for practical purposes, no problem.

For information on those physical properties of superfine particles which have not been touched on in this chapter, such as thermal

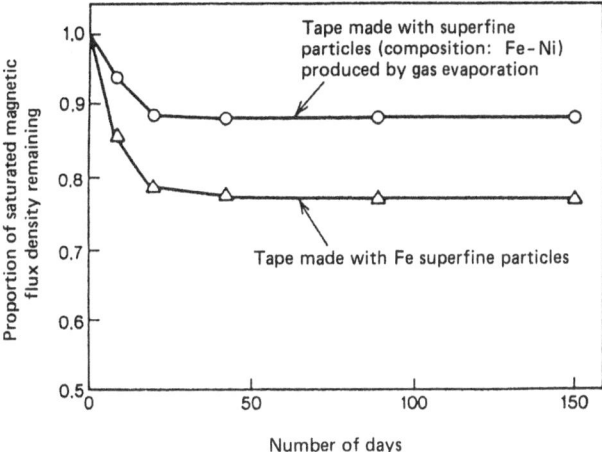

Fig. 2.10 Comparison of changes over time in the properties of computer tape.

properties and superconductivity, please refer to the papers published by researchers in the relevant areas [34].

References

1. S. Iijima: Jpn. J. Appl. Phys. 23, p. L347 (1984)
2. K. Kimoto: Nihon Kesshou Gakkaishi, 10, p. 160 (1968) (in Japanese)
3. Hachiya, Ueda: Ouyou Butsuri, 42, p. 1067 (1973) (in Japanese)
4. R. Uyeda: Nihon Kesshou Gakkaishi, 16, p. 19 (1974) (in Japanese)
5. K. Kimoto: Nihon Kinzoku Gakkaihou, 12, p. 107 (1973) (in Japanese)
6. R. Uyeda: Nihon Kinzoku Gakkaihou, 17, p. 403 (1978) (in Japanese)
7. S. Iijima: Shin Gijutsu Kaihatsu Jigyoudan Souzou Kagaku Gijutsu Suishin Jigyou Hayashi Cho-biryushi Project Research Report, p. 3 (1984), p. 10 (1985) (in Japanese)
8. S. Iijima, T. Ichihashi: Jpn. J. Appl. Phys., 24, p. L125 (1985)
9. S. Iijima: Tokushuu Cho-biryushi Nihon no Kagaku to Gijutsu, 25, 227, p. 28, Nihon Kagaku Gijutsu Shinkou Zaidan (1984) (in Japanese)
10. S. Iijima: Ouyou Butsuri, 54, p. 926 (1986) (in Japanese)
11. R. Uyeda: Aerosol Kenkyu, 54, p. 99 (1986) (in Japanese)
12. R. Uyeda: Ultramicroscopy, 20, p. 29 (1986)
13. C. Hayashi: Cho-biryushi Kagaku Sousetsu, 48, p. 1, Gakkai Shuppan Center (1985) (in Japanese)
14. A. Tasaki: Jpn. J. Appl. Phys., 4, p. 707 (1965)
15. A. Tasaki: "Cho-biryushi Kotai Butsuri Bessatsu Tokushugou", p. 70, Agne Gijutsu Center (1975) (in Japanese)
16. Imanishi, Inoue, Ono, Miyatake: National Technical Report, 25, p. 153 (1979)
17. Inoue, Imanishi, Ono, Miyatake: Television Gakkai Gijutsu Houkoku, VR34-1 (1979)
18. M. Takagi: Kinzoku, 53, 5, p. 33 (1983) (in Japanese)
19. C.R.M. Wronski: Br. J. Appl. Phys., 18, p. 1731 (1967)

20. Hayashi, Masuda: Funtai Funmatsu Yakin Kyoukai 1974 Fall Taikai Abstracts, p. 42 (1974) (in Japanese)
21. M. Sahashi: Nihon Kinzoku Gakkai "Kinzoku Cho-biryushi no Seisaku kara Ouyou made", Symposium paper p. 26 (1975) (in Japanese)
22. S. Iwama, K. Hayakawa: Jpn. J. Appl. Phys., 20, p. 335 (1981)
23. E. Kamijou: Shinkuu, 19, p. 212 (1976) (in Japanese)
24. E. Kamijou: Cho-biryushi no Jitsuyouka Gijutsu, p. 75, CMC (1984) (in Japanese)
25. K. Itou: Shinkuu, 16, p. 163 (1973) (in Japanese)
26. K. Itou: Kinzoku, 45, 7, p. 65 (1975) (in Japanese)
27. Ichikawa, Hayashi: "Cho-biryushi Kotai Butsuri Bessatsu Tokushugou", p. 117, Agne Gijutsu Center (1984) (in Japanese)
28. Y. Saitou: Tokushuu Cho-biryushi Nihon no Kagaku to Gijutsu, 25, 227, p. 59, Nihon Kagaku Gijutsu Shinkou Zaidan (1984) (in Japanese)
29. Miyoshi, Saitou: Chemical Society of Japan 49th Spring Nenkai paper, 2Q06 (1984) (in Japanese)
30. T. Hayashi: Shin Gijutsu Kaihatsu Jigyoudan Souzou Kagaku Gijutsu Suishin Jigyou Hayashi Cho-biryushi Project Research Report, p. 3 (1983) (in Japanese)
31. Oda, Hayashi: 23rd Funtai nikansuru Touronkai Abstracts, p. 122 (1985) (in Japanese)
32. A. Tasaki: "Cho-biryushi Kotai Butsuri Bessatsu Tokushugou", p. 111, Agne Gijutsu Center (1984) (in Japanese)
33. Nakamura, Abe: Kinzoku Hyoumen Gijutsu Kyoukai 65th Gakujutsu Kouen Taikai Shishuu, p. 26 (1975) (in Japanese)
34. The following are useful references dealing with the physical properties of ultrafine particles:
    Kotai Butsuri Bessatsu Tokushugou Cho-biryushi, Agne Gijutsu Center (March 1984) (in Japanese)
    "Kagaku Sousetsu No. 48 Cho-biryushi – Kagaku to Ouyou", Gakkai Shuppan Center (September 1985) (in Japanese)

Additional References

35. Chikara Hayashi: "Ultrafine particles", Phys. Today, Dec., pp. 1–8 (1987)
36. K. Hatanaka, M. Kaito, M. Umehara, S. Kashu, C. Hayashi: "Preparation of superconducting thick films of Y–Ba–Cu–O by gas deposition of ultrafine powder, Proceedings of the 1st International Symposium on Superconductivity (ISS'88), pp. 341–345 (1988)
37. C. Hayashi, R. Ueda, A. Tazaki (eds.): "Cho-biryushi – Sozo, Kagaku Gijutsu", Mita Shuppankai (1988) (in Japanese)
38. M. Koishi (ed.): "Cho-biryushi Ouyou Kaihatsu Handbook", Science Forum, (1989) (in Japanese)
39. Chemical Engineering Tokushu: "Hakumaku, Cho-biryushi Zairyo eno Kitai", vol. 33, no. 2, pp. 113–153 (1988) (in Japanese)

Chapter 3

Chemistry of Superfine Particles

How should we define the chemical properties of superfine particles? Chemistry could be described as the science of the "transformation of matter". A transformation is deemed to have occurred when a comparison is made between two states in which a certain type of matter exists, and a clear difference is found between them. Superfine particles undergo major transformation in their state of existence due to their interaction with the surrounding environment. These changes are most striking when the surrounding environment is a liquid.

In this chapter, we shall look first at the phenomenon of adsorption, which is the chief cause of these changes, and which occurs on the surface of fine particles in liquid. Next, we shall deal with the closely related subject of the dispersal and flocculation of fine particles in liquid. On the way, we shall touch on a subject of great importance in industrial applications of fine particles: the rheology of liquids in fine particle dispersals, or in other words, the deformation and flow characteristics of particle disperse system liquids. Finally, we shall discuss fine particle disperse system gels. The latter are attracting much attention in connection with the sol–gel method, a new technique for material production.

Introduction

.viaterials, as well as having structural regularity at the atomic or molecular level – generally known as crystal structure – have structural regularity at a much higher size level, and this is known as the organization, or microstructure, of the material. Recent advances in science and technology are resulting in industrial materials which are much less bulky, perform much better, and are much more reliable. The need for such materials is particularly acute in the area of electronics. To meet this need, finer and finer particles are being used as the raw material in the production of materials such as ceramics, manufactured by forming and then sintering a fine powder, sintered metallic materials, and pigments and paints, which are disperse systems of powder particles. The use of superfine particles in these materials is now commonplace, but there is still no precise definition of the term "superfine particle", and there is no clear dividing line between a "superfine particle" and a "fine particle". The term "superfine particle" is therefore used by different scientists to refer to particles of different sizes. Broadly speaking, however, a consensus seems to have been reached that superfine particles are sub-micrometer particles, or in other words, particles with a diameter of less than 1 μm. This is exactly the same as a colloidal particle, which has a diameter of between 10^{-1} and 10^{-3} μm, which gives a stable fine particle disperse system.

In 1861, T. Graham coined the term "colloid" for a continuous substance with a low diffusion speed. In the years that followed, many studies were carried out on substances of this type. The most famous of these resulted in the discovery that when a beam of light is shone into a colloidal liquid, the particles or molecules (polymers) in the light beam cause light scattering, so that the light beam is seen to sparkle. The first experiment on colloid light scattering was carried out in 1802 by Richter. In 1869, a more detailed study of colloid light scattering was carried out by Tyndall, and this phenomenon was later named the Tyndall effect after him. The colloid discovered by T. Graham was a polymer solution, but it is known that superfine particles other than this, stably dispersed in a liquid, show the same behavior as polymer solutions. Superfine particles stably dispersed in a liquid are therefore termed a colloid disperse system, and viscous solutions such as colloid disperse systems and polymer solutions are referred to by the blanket term of sols. Gels, furthermore, resemble sols, and although a gel is more viscous than a sol, the difference between the two is not clearly defined. In general, however, the substances with fluidity are called sols, and those without are called gels.

In non-technical terms, a sol can be thought of as a solid with fluidity, and a gel as a liquid without fluidity. In recent years, a good deal of interest has been generated by the sol–gel method, a new technique used in fine ceramics production, in which a sol is used instead of a powder. The advantage of producing ceramics by the sol–gel method instead of the conventional method involving powdered materials is that the flocculation of the particles can be controlled. For example, in the production of ceramics by the pressure-molding of powder, if the powder forms lumps of different shapes and sizes, this will lead to faults in the final product. Similarly, when carrying out ceramic production using methods such as slip casting or tape casting, slip which contains lumps can also cause faults in the final product. If the material used is a sol with stably dispersed particles, flocculation is even and monodispersed, making it possible to achieve close to perfect sintering.

Generally, when superfine particles are used as an industrial material, they are dispersed in a liquid, and used in a sol or gel state. This means that it is important, even vital, to know how superfine particles behave in a liquid. In this chapter we shall describe the physical and chemical behavior of superfine particles in liquid, with particular reference to adsorption, dispersion and flocculation, the rheology of sols, and gels.

3.2 Adsorption

Adsorption is one of the interfacial phenomena which occur between different phases which have come into contact. It involves the retention of the adsorbate in a very thin layer on the surface or at the boundary of the solid or liquid which is the adsorbent. There are two sorts of adsorption: when the adsorbent and the adsorbate are linked by comparatively weak physical forces such as van der Waals' forces, the phenomenon is known as physical adsorption, while if adsorbent and adsorbate are linked strongly by a chemical bond or a bond of a similar energy level, it is known as chemical adsorption. Many different adsorbent–adsorbate combinations are possible, but ceramic particles are the only adsorbent we shall deal with here. We shall examine the way in which the adsorbent adsorbs the adsorbate from the solution in three cases: (a) when the adsorbate is a non-electrolyte [1], (b) when the adsorbate is an electrolyte [2], and (c) when the adsorbate is a macromolecular substance [1]. We shall give a brief outline of the adsorption process in each case.

3.2.1 Adsorption of Non-electrolytes

A non-electrolyte consists of atoms which do not have an electric charge, so it is adsorbed at the particle surface by hydrogen bonding, van der Waals' forces, and weak electrostatic attraction from dipoles. Of these, hydrogen bonding plays the most important role. Consider the following example: Fig. 3.1 shows alcohol, amide, and ether being adsorbed at the surface of a silica particle at a low pH. As shown in the diagram, the π electrons of the electronegative atoms of alcohol, amide and ether are adsorbed by forming hydrogen bonds with the hydrogen atoms of the silanol group on the surface of the silica particle. When the silica particle is dehydrated by heating, a single molecule of water is removed from the adjacent silanol group and forms a siloxane bond on the surface of the silica particle. The state of the surface is therefore very different from the pre-dehydration silanol group. This means that the adsorption of

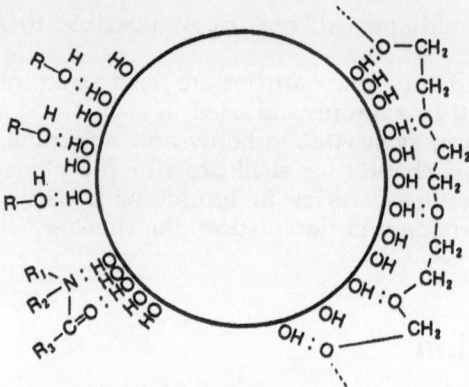

Fig. 3.1 Alcohol, amide and ethyl molecules being adsorbed at the surface of a silica particle at a low pH.

substances at the surface is also different. For example, the amount of methyl red adsorbed into a silica particle, the surface of which has been particularly dehydrated by heating, from a benzene solution of methyl red, is less than the amount adsorbed by a silica particle not subjected to dehydration by heating. This shows that the methyl red is adsorbed by the surface silanol groups, but not by the siloxane bonds. The same phenomenon has been observed regarding molecules of water. That is to say, the water molecules are also adsorbed by hydrating the silanol groups at the surface of the silica particle. In the case of alcohol, the alcohol molecules are adsorbed by the O–H hydrogen bond formed between the O on the silica particle surfaces and the H of the OH group of the alcohol. As can be seen from the diagram, this is physical adsorption: the bonding is weak because adsorption results from one hydrogen bond with the alcohol molecule. The adsorption of polyethylene oxide macromolecules works in the same way, due to O–H hydrogen bonds, but so many bonds are contributed by adsorption, and the adsorption is so strong, that this could in fact be regarded as chemical adsorption.

Thus adsorption is strongly affected by the properties of the adsorbate molecules as well as the surface properties of the adsorbent particles. Even for the same adsorbate, the amount adsorbed varies according to the solvent. For example, if a straight-chain fatty acid is caused to be adsorbed at the surface of a silica particle from a benzene or a hexane solution, the amount adsorbed is greater when a hexane solution is used than when a benzene solution is used. This is because, in benzene, the fatty acid becomes a dimer, due to hydrogen bonding, and, at a state of equilibrium, there are hardly any monomers with hydrogen bonding capacity. The adsorption of a non-electrolyte from an aqueous solution is strongly affected by pH. At a high pH, the surface of the silica particles takes on a negative electric charge, and the hydrated counter ions move close to them, interfering with the formation of hydrogen bonds in that area. Figure 3.2 is a schematic diagram of this process. The diagram shows how the hydrated sodium ions are adsorbed at those parts of the surface which have taken on a negative electric charge, thus preventing the adsorption of polyethylene oxide or alcohol molecules in the surrounding area. When the pH rises and a negative electrical charge is added, this trend gets stronger.

3.2.2 Adsorption of Electrolytes

Unlike non-electrolytes, electrolytes in solution exist in the form of ions, and Coulomb force is the dominant factor in their adsorption.

In order to balance the electric charge which naturally forms on their surfaces, the particles need an opposing electric force. The ions in the solution which have an opposing electrical charge are termed counter-ions. In clay, the surface charge is almost always negative, and is, as a rule, neutralized by the positive ions of the alkali or alkali earth. When the adsorbed ions are Ca^{2+}, the clay is generally termed Ca-clay. For example, when Ca^- clay particles are suspended in water, the Ca^{2+} ions move a little further away from the surface, and form what is known as a diffusion layer of counter-ions. This layer can be divided into two sub-layers: the one on the inside, which is strongly adsorbed at the particle

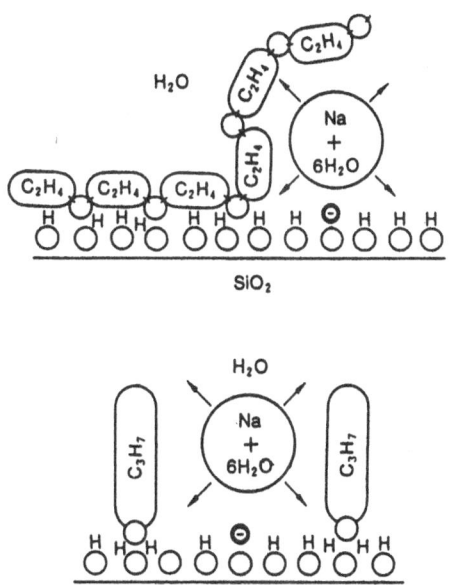

Fig. 3.2 Prevention of molecule adsorption by large counter-ions.

surface is called the Stern layer, and the more weakly adsorbed layer on the outside is called the Gouy-Chapmann layer. The Stern layer is fixed, and consists of some of the adsorbed counter-ions: it neutralizes some of the charge on the surface of the particle. As the majority of the positive ions remaining are drawn further away from the electrically charged particles by the force of attraction produced by the electric charge on the surface of the particle, they gradually form a diffused layer with a different diffusion to that of the liquid, in which negative and positive ions exist in equal numbers. The double layer thus formed around the charged particles is termed an electric double layer. The electric potential of this double layer falls steeply inside the Stern layer, and gradually inside the Gouy–Chapmann layer. The particles dispersed in the solution move in the electric field, and some of the liquid has now been adsorbed by the particles and shifts with the particles. The electric potential at the interface between the liquid adsorbed by the particles and the rest of the liquid is called the zeta electric potential. The zeta electric potential is an index of the thickness of the double layer, and can be determined by means of an electrophoresis experiment.

The two-layer distribution of the counter ions can be expressed quantitatively by electric potential. Now, let us carry out the calculation on the assumption that a particle of a substance such as Ca-clay is a strong electrolyte. If we take the particle surface as the starting point, and calculate the electrical potential Ψ at the desired distance x on the solution side, Ψ can be expressed approximately by:

$$\Psi = \Psi_0 \exp(-\kappa x) \tag{3.1}$$

However,

$$\kappa = \left(\frac{2\,e^2 n_0 Z^2}{\epsilon k T}\right)^{\frac{1}{2}} = \left(\frac{2\,e^2 N_A C Z^2}{\epsilon k T}\right)^{\frac{1}{2}} \tag{3.2}$$

Here, when $x \to \infty$, it is assumed that $\Psi = 0$ and the electric potential of the particle surface is Ψ_0. In actual fact, the zeta electric potential ζ to be obtained experimentally can be supposed to be Ψ_0. ϵ is the dielectric constant of the solution, e is the electric charge of the electron, n_0 is the ion concentration, Z is the valence, k is Boltzmann's constant, N_A is Avogadro's number, C is the mole concentration (mol/cm^3) of a strong electrolyte, and T is the absolute temperature.

Eq. 3.1 indicates the form of the index function Ψ. This means that κ is proportional to the thickness of the double layer. In this formula, $1/\kappa$ can be called the thickness of the double layer. From Eq. 3.2 it can be seen that $1/\kappa$ is inversely proportional to C and Z^2, so it is clear that the higher the electrolytic density, and the higher the valence, the thinner the double layer will be.

Figure 3.3 shows how this happens in the case of kaolin, for example. When the OH$^-$ ion concentration is sufficiently high, the edges of the lamellar kaolin particles become negatively charged. Figure 3.3(a) shows the state of the double layer which forms around a negatively charged particle. If, at this point, there are enough monovalent positive ions – such as Na$^+$ – to form a diffused layer, the layer then thickens as shown in Fig. 3.3(b), and Ψ slowly decreases. However, if the concentration of the Na$^+$ ions is then increased even further, the number of effective counter-ions increases. Therefore, as shown in Figure 3.3(c), Ψ decreases sharply, and the double layer becomes thinner. Also, even when the number of counter-ions is just right, and a double layer of sufficient thickness has been formed, then the electric charge of the counter-ions is increased, Ψ naturally decreases sharply, as shown in Fig. 3.3(d), and the double layer becomes thinner.

Similarly, with oxides, the type and number of counter-ions determines the thickness of the double layer. As shown in Fig. 3.4, in an aqueous solution, the surface of a particle of an oxide such as quartz, aluminum oxide, and titanic oxide becomes positively, neutrally or negatively charged, depending on pH. In a state where the surface charge is positive, the effective counter-ions are large negative ions such as Cl$^-$ and NO^{3-}, and when the surface charge is negative, monovalent alkali ions such as Na$^+$, or NH^{4+} are the effective counter-ions.

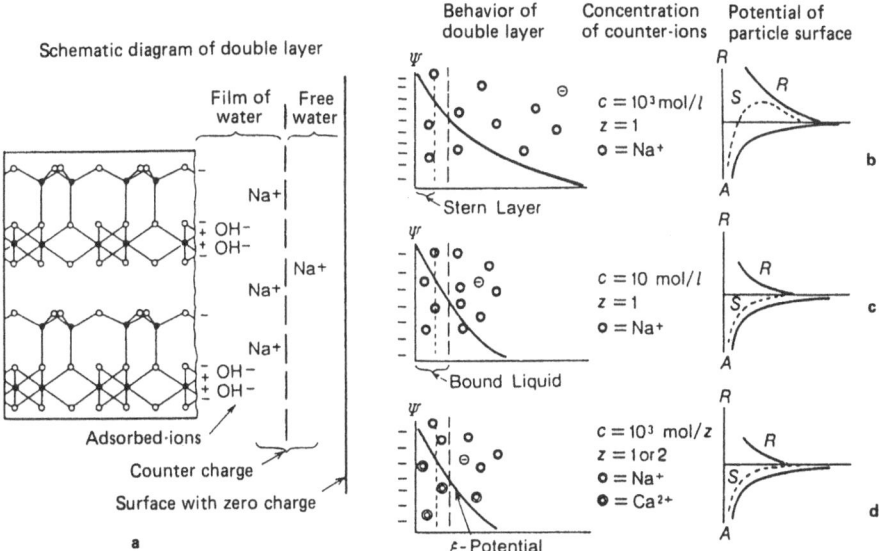

Fig. 3.3 Relationship between thickness of electric double layer and concentration of counter-ions. S, surface potential; R, repulsion; A, attraction.

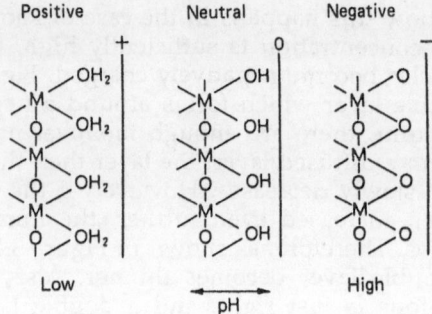

Fig. 3.4 pH-related changes in the charge on the surface of an oxide particle.

3.2.3 Adsorption of Macromolecules

Some macromolecules are electrolytes and some are non-electrolytes. The mechanism of adsorption of non-electrolyte macromolecules is basically as described in Sect. 3.2.1, and the mechanism of adsorption of electrolyte macromolecules is basically as described in Sect. 3.2.2. However, there is a difference in that macromolecules contain many bonds, and, because they are large, steric hindrance occurs. This is why macromolecules are covered here in a separate section. The adsorption of macromolecules is strongly influenced by factors such as the state of the surface of the adsorbent, and the state of the adsorbate. This is clearly shown in the co-adsorption of polyvinyl alcohol (PVA) and cation surface active agents. Figure 3.5 shows how adsorption varies according to differences in pH. At low pH, the surface of the silica particle consists of silanol radicals, and due to the hydrogen bonding of these silanol radicals with the OH radicals of the PVA, the PVA turns the silica particle surface inside out so that the hydrophobic radicals face outwards, and the hydrophobic radicals of the cation surface active agent form hydrophobic bonds with the hydrophobic radicals of the PVA. At high pH, the silica surface has a negative electrical charge, so in order to neutralize this, the cation surface active agent forms a micelle, which then adsorbs the PVA by hydrophobic bonding. At a neutral pH, some parts of the silica particle surface have a negative electrical charge, and the cation surface active agent is adsorbed in these parts, while the PVA is adsorbed in the remaining, silanol parts.

Hostetler and Swanson carried out a detailed experiment on the adsorption of polyethylene imines (PEI) into silica gel, using a silica gel–PEI–water system [3]. The properties of the silica gel used in the experiment are shown in Table 3.1.

Figure 3.6(a) shows the variations in the ionization rate of the PEI when the pH is varied: it can be seen that ionization progresses further as the pH gets lower. Figure 3.6(b) shows the light scattering rate for the

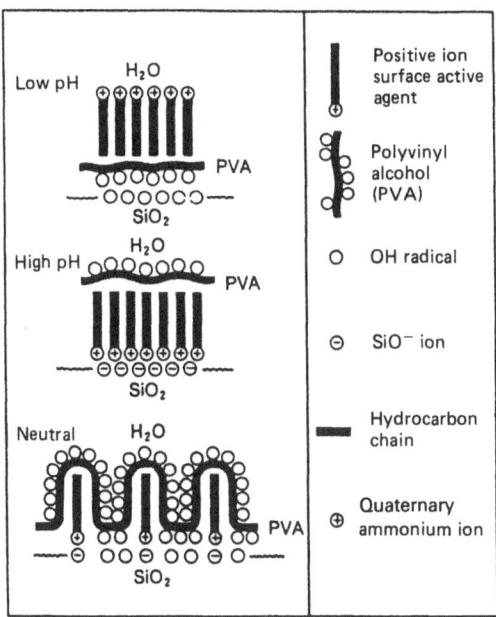

Fig. 3.5 Co-adsorption of polyvinyl alcohol (PVA) and cation active surface agent at the surface of a silica particle at different pH values.

Table 3.1 Properties of silica gel used in the adsorption experiment

BET surface area (m^2/g)	Cumulative pore surface area (m^2/g)		Total pore volume (cm^3/g)		Range of most frequent pore sizes (nm)
	NAI	MIP	NAI	MIP	
282	257	350	0.90	0.93	5–12

NAI: Data obtained from nitrogen gas adsorption
MIP: Data obtained from mercury pressure porosimeter

PEI solution at various pH values: it can be seen that as the pH rises, the light scattering rate falls. This means that at a high pH, the PEI is contracted and there is little scattering, while at a low pH, it is expanded and there is a good deal of scattering. If the PEI adsorption depends mainly on molecule size, it is predicted that a good deal of the PEI will be adsorbed at the high pH, in which case the PEI contracts, and the amount adsorbed will increase as the pH rises. However, as shown in Fig. 3.7, actual PEI adsorption is more idiosyncratic. Its adsorption characteristics are understood to be as follows. In solutions with a pH

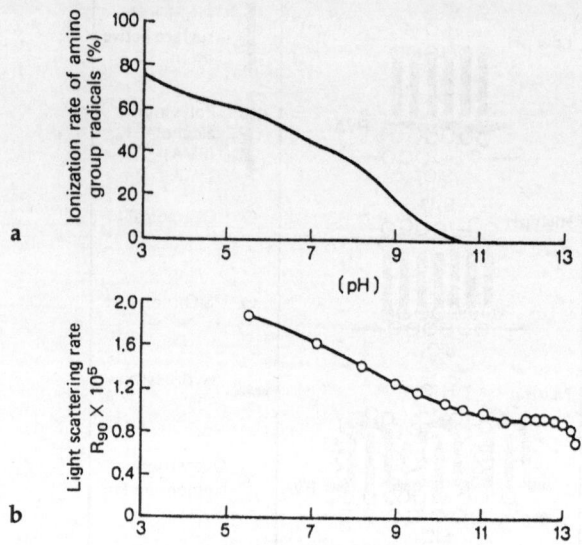

Fig. 3.6 pH variations for **a** PEI ionization rate and **b** light scattering rate (measure in a 0.109 N NaCl solution).

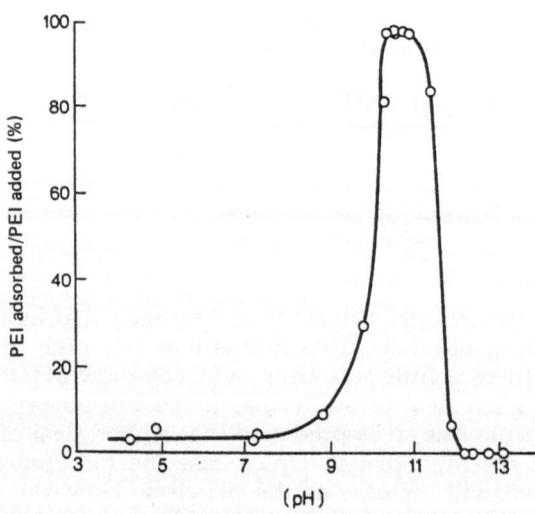

Fig. 3.7 Amount of PEI adsorbed into silica particle at different pH values.

of 9 or under, the PEI ions expand, so nothing can get into the pores, and the amount adsorbed is small. For pH values between 10 and 12, PEI takes on a coil shape, so the pores are accessible, and some of the polymer chains are ionized in the form $(C_2H_4)_2NH_2^+$, and are adsorbed into the pores. On the other hand, at a pH value of 12 or over, it is thought that a considerable part of the negatively charged silica particle surface is hydrated, preventing PEI from approaching and being adsorbed.

3.3 Particle Dispersion

When superfine particles are dispersed by an ordinary dispersant, they usually settle with the passage of time, except in special cases. The settling time, or settling speed, is determined by factors such as the viscosity and specific gravity of the dispersant, particle size, and specific gravity of the dispersion medium. However, when the particles are as small as colloids – between 1 and 500 nm – settling is prevented by factors such as the Brownian motion of the dispersion medium, and the particles remain stable in a dispersed state. This stable dispersed state is termed a colloid disperse system, or sol. When the dispersion medium is water, the result is known as a hydrosol, and when an organic solvent is used, it is known as an organosol.

Generally, even when fine particles of colloid dimensions are formed, flocculation naturally occurs unless steps are taken to prevent it. Even when the primary particles are of colloid dimensions, if they form large lumps a stable dispersed state cannot be achieved, and settling occurs. In order to make the particles repel each other and prevent flocculation, it is important to produce the electric double layer described in Sect. 3.2.2. That is to say, a double layer needs to be formed around the particles, so that the particles are kept apart by electrical repulsion. This can also be done by causing the molecules of the surface active agent surrounding the particles to adsorb, and form micelles, and then using the steric hindrance they provide.

Generally speaking, there are three methods of producing a sol. These are as follows: (a) the peptization method, in which a peptization agent is added to the lumps formed by flocculation of colloidal particles, to peptize them; (b) the comminution method, in which large particles are comminuted to colloid size; and (c) the agglomeration method, in which the ions and molecules are made to agglomerate and grow to form particles of colloidal size.

3.3.1 Peptization Method

Stable disperse systems are produced by using a peptization agent to partially cancel the attractive force which acts between particles and causes flocculation. Attractive forces are also generated by polarization, and by Coulomb interactions, but the attractive forces most important in peptization are van der Waals' forces. As we stated earlier, almost all ceramic particles form lumps when in a suspension, unless steps are taken to set up repulsion between the particles. Peptization is the operation of preventing flocculation and causing the particles to disperse stably in the solution. Peptization is carried out by creating an electric double layer around the particles. When the particles are in water, they are neutral, no matter what electrical charge the particle surface has. This is due to the generation of the electric double layer. The electric double layer is affected by the charge of the particle surface, and this charge in turn depends on the structure and composition of the particles.

Clay, a natural substance which consists of fine particles, is a mixture of minerals such as kaolin, montmorillonite and illite, plus quartz and organic substances. Usually, clay is about 50% kaolin, and the rest is montmorillonite, illite, mica, quartz and organic substances.

Kaolin mineral particles take the form of hexagonal disks ranging in size from 2 μm to several hundred micrometers across, and consist of layers of gibbsite ($Al(OH)_3$) and layers of silica. The layers are arranged in a certain order, and bonded by hydrogen bonds. Theoretically, each layer could extend infinitely within the layer surface (in crystallographic terms, in the a–b direction), but in practice, warping occurs between the layers due to small differences in size between the basic units of the gibbsite layer and the silica layer, resulting in small lamellar particles with an unsaturated electrical charge at the edges. In the kaolin structure, Al^{3+} ions break away easily from the gibbsite layer, and Si^{4+} ions from the silica layer, producing a negative electrical charge on the surface of the lamellar particles. At the edges of the lamellar particles, the array of large O^{2-} ions and OH^- ions enclosing the small Al^{3+} ions and Si^{4+} ions vary according to pH, and can thereby be given a positive, neutral or negative electrical charge. Figure 3.8 is a schematic diagram of the variation in the electrical charge of the clay particles according to the variation in pH. This diagram also shows the flocculation of the particles for each electrical charge.

Montmorillonite, mica and illite are formed when the positive ions in the gibbsite or the silica layer of the pyrophyllite – a three-layered mineral – are replaced by positive ions with a lower valence. Pyrophyllite has a sandwich structure consisting of a gibbsite layer between two layers of silica. In pyrophyllite, the Al^{3+} ions in the gibbsite layer are replaced by Mg^{2+} ions, so the whole of the particle surface becomes negatively charged. This electrical charge is compensated by the positive ions which have entered the adjacent layer, so the properties of

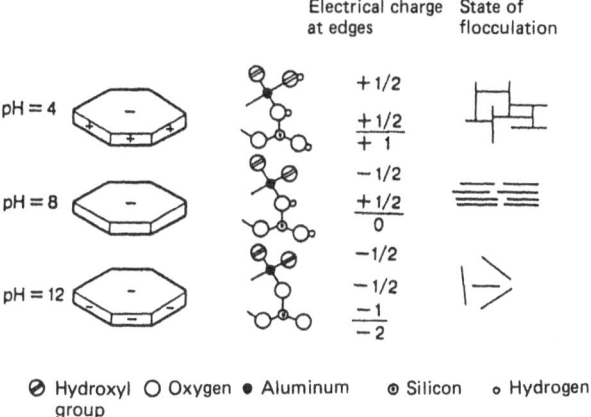

Fig. 3.8 Variations in the electrical charge of clay particles according to variations in pH.

montmorillonite are strongly influenced by the properties of the positive ions.

Mica and illite are produced in a similar way by the replacement of some of the Si^{4+} ions in the silica layer of the pyrophyllite by Al^{3+} ions. In mica, a quarter of the Si^{4+} ions are replaced, and in illite, a sixth. The compensating ions in muscovite are K^+, while in illite they are Ca^{2+}, Mg^{2+} and H_3O^+ ions as well as K^+ ions. In muscovite, there are K^+ ions at the edges of the layers, and these become the bonding ions and form mica. The reason why mica flakes off in fine layers is that the bonding has been broken where these K^+ ions are. Also, in illite, the accumulation at right angles to the surface is irregular, and this is why it breaks up into small scale-like pieces. In mica, illite, and montmorillonite, as in kaolin, the edges can be given a negative, neutral or positive electrical charge by changing the pH, as shown in Fig. 3.8.

In most oxides, there is a small positive ion enclosed by large O^{2-} ions, and this structure is heavily dependent on the ratio between the radius of the positive ion and the radius of the O^{2-} ion. Oxide particles suspended in water generally hydrate. The degree of hydration varies according to the oxide, but in oxides such as MgO and CaO, they become complete hydroxides. As with clay, the degree of hydration in these cases depends on the pH. For each oxide there is a particular pH at which the hydrated surface becomes neutral.

3.3.2 Comminution Method

The comminution method consists of comminuting large particles down to colloid dimensions, but in order to prevent the comminuted particles from re-aggregating, they must also be treated to set up repulsion between the particles. We shall now describe the comminution method of particle dispersion, taking a magnetic fluid as an example of a typical dispersed system manufactured by comminution. A magnetic fluid is a fluid in which superfine magnetic particles are stably dispersed, and in which centrifugation, magnetic fields and other factors do not normally cause flocculation. Papell carried out extended ball-mill comminution of magnetite with 30 μm particles in oleic acid, and synthesized a magnetic fluid in which 10 nm magnetite particles were stably dispersed [4]. In this example, the oleic acid contributes to the prevention of flocculation, instead of a electric double layer. That is to say, the adsorbed oleic acid encloses the magnetite particles, and this provides repulsion between the particles. This force of repulsion is due to the effect of steric hindrance in the layer of adsorbed oleic acid which has formed on the surface of the particles. Since then, many researchers have produced ferrite-system magnetic fluids by a variety of methods, but all were alike in that they required a micelle state in which the oleic acid or oleic acid ions surrounded the particles. A layer of adsorbed molecules, with thickness δ, surrounds each magnetic particle, which has radius r. Rosensweig has carried out a theoretical study of the repulsion energy generated by steric hindrance in oleic acid [5]. If we base our assumptions on a theoretical model, such as Fig. 3.9, of the strongly magnetic particles into which the oleic acid has been adsorbed, the repulsion energy V can be given in the following equation

$$V = 2\pi r^2 NkT \left\{ 2 - \frac{(h+2)}{\delta/r} \ln \left(\frac{1+\delta/r}{1+h/2} \right) - \frac{h}{\delta/r} \right\} \qquad (3.3)$$

Here, N is the number of adsorbed molecules per unit volume, δ is the thickness of the adsorption layer, k is Boltzmann's constant, and T is the absolute temperature. h is the function for the distance R between the particles relative to the particle radius r and is expressed in the following way: $h=R/r-2$. That is to say, when the particles are touching, $h=0$, and

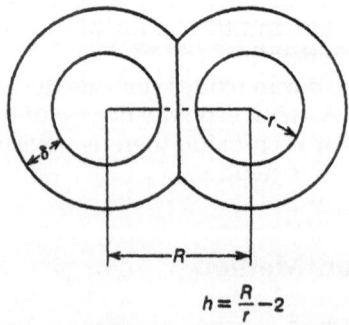

$$h = \frac{R}{r} - 2$$

Fig. 3.9 Model of particles in a magnetic fluid. r, Radius of magnetic particles; δ, thickness of adsorbed molecular layer.

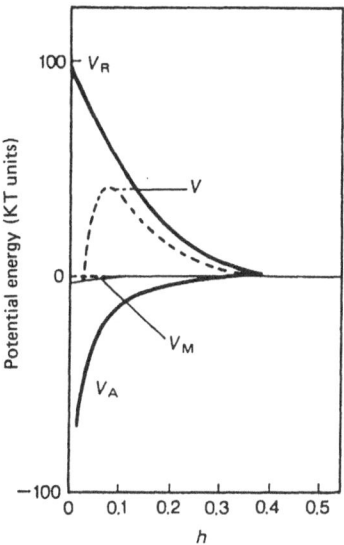

Fig. 3.10 Potential graph for 10 nm magnetic particle. V_R, repulsive potential due to steric hindrance; V_A, Attractive potential due to van der Waals' forces, V_M, Magnetic attraction.

the further apart the particles are, the greater the value of h. Assuming that the forces acting on the particles are the van der Waals' forces V_A, the force of magnetic attraction V_M, and the force of repulsion V_R due to steric hindrance, if we calculate in terms of the repulsion energy V acting on a 10 nm magnetic particle, and with the various forces as a function of h, the result will be as shown in Fig. 3.10 [6]. However, the thickness of the adsorbed layer has been calculated as 1 nm, and the number of molecules in the adsorbed layer has been calculated as 3.3×10^{14}. As shown in the diagram, there is a large peak of potential energy around the particle. To make the particles flocculate, a force of attraction is needed which is sufficient to make the particles cross their potential-boundaries and join together. This is why the result is a stable disperse system.

3.3.3 Agglomeration Method

Many stable colloid disperse systems are produced by hydrolysis of inorganic and organic salt solutions. The hydrolysis reaction causes the ions in the solution to grow into colloidal size particles by agglomeration. For example, if the molecular disperse system tetra-ethoxysilane is refluxed with water, colloid sized silica particles are generated, and a silica sol is obtained in which these particles are stably dispersed. These particles are repulsed by their electric double layer. The counter-ions at this stage are thought to be H_3O^+. Experiments have also been performed in which slow hydrolysis of a metal salt was carried out at a temperature between room temperature and 250°C, in acidic conditions, for a period between several hours and a few weeks, and monodispersed particles – spheroidal, tetragonal and other shapes – were obtained of hydrated oxides such as TiO_2, Al_2O_3, Fe_2O_3 and ThO_2, sulfides such as CdS, and carbonates such as $CdCO_3$ [7]. In this method, metal ions in solution are precipitated *in situ* as colloidal particles, by means of the catalytic effect of the negative ions which are also present. This method makes it possible to generate particles with extremely good monodispersion. In the main, solutions of inorganic salts, such as sulfates, chlorides, and nitrates have been used, but in recent years, researchers at Massachusetts Institute of Technology have synthesized monodispersed spheroidal particles of SiO_2 and TiO_2 using metal alkoxides as starting materials [8].

3.4 Particle Flocculation

Particles with a fully thickened electric double layer are stably dispersed in a solution, and deflocculated. Let us consider what causes particles in a solution to flocculate. Particles suspended in a solution are universally acted upon by van der Waals' forces. The factor which prevents them from flocculating, due to the van der Waals' forces, and which produces stable dispersal is the effect of the electric double layer. Therefore, particle flocculation must be considered in the light of the interaction between the electric double layer and the van der Waals' forces acting between the particles.

E_V, the mutual potential energy generated by the van der Waals' forces between two particles with radius r, can be expressed by the following equation

$$E_V = -\frac{A}{12} \cdot \frac{r}{l} \tag{3.4}$$

Here, l is the shortest distance between the particles, r is the particle radius, and A is a constant.

E_0, the potential energy for mutual interaction of the electric double layer, can be expressed approximately by the following equation

$$E_0 \sim \frac{\epsilon r \Psi_0{}^2}{2} \exp(-\kappa l) \tag{3.5}$$

Here, ϵ is the dielectric constant of the solution, Ψ_0 is the electric potential of the particle surface, and κ is expressed by Eq. (3.2).

Accordingly, E, the total mutal interaction energy between the two particles, is expressed by the following equation

$$E = E_V + E_0 = \frac{\epsilon r \Psi_0^2}{2} \exp(-\kappa l) - \frac{A}{12} \cdot \frac{r}{l} \qquad (3.6)$$

If we express E, E_V, and E_0 as the function l – the distance between the particles – the result is as shown in Fig. 3.11. When κ is small, E is at its highest, and there is an energy barrier hindering flocculation. When κ is high, there is no such energy barrier. What happens in the former case is called slow flocculation, and the latter, rapid flocculation. The electrolyte concentration which gives a value for κ such that $E_{max}=0$ is termed the critical flocculation concentration, or the coagulation value.

Accordingly, from Eq. (3.6) and from the facts that $E=0$ and $(dE/dl)=0$, the critical flocculation concentration C_f can be calculated from the following equation

$$C_f = \frac{16\epsilon^3 kT}{N_A e^4 A^2} \cdot \frac{\Psi_0^4}{Z^2} \propto \frac{1}{Z^2} \qquad (3.7)$$

This shows that the critical flocculation concentration C_f is inversely proportional to the valence Z of the counter-ions. Many studies have already been carried out on the influence of electrolytes on the flocculation of colloid particles, and it is now known that the minimum electrolyte concentration necessary for flocculation – that is, the critical flocculation concentration – does not depend on the type of ion, but only on its

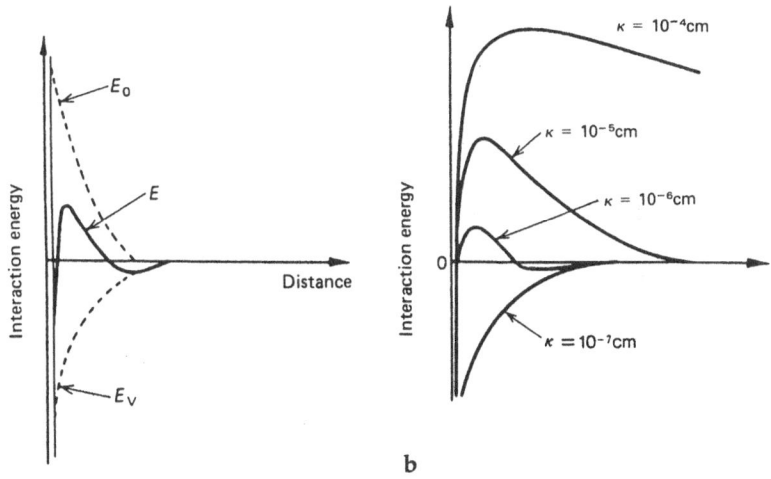

Fig. 3.11 Energy of interaction between particles. **a** Qualitative curve, and **b** variations in E curve due to κ.

valence. This is known as the Schulze–Hardy rule. It can be expressed quantitatively as follows:

$$C_f \propto \frac{1}{Z^6} \tag{3.8}$$

Equation (3.7) does not take this form, because an approximate formula was used in Eq. (3.5), but when the calculation is done more accurately, the above equation is obtained. This equation shows that the flocculation of colloid particles can be accounted for by the interaction between the van der Waals' forces and the electric double layer.

3.5 Rheology

Rheology is the "science of deformation and flow in matter" [9]. As we have already stated, fine particles behave very differently to the original solid, because of the huge size difference. The rheology of colloids – disperse systems of fine particles of 1 μm and under in liquid – is particularly interesting from both the scientific and the industrial viewpoint. In this section, we shall look at viscosity, the most important rheological property of a colloid disperse system.

As shown in Fig. 3.12, two flat, parallel plates separated by distance h are set up, and the space between them is filled with fluid. The lower

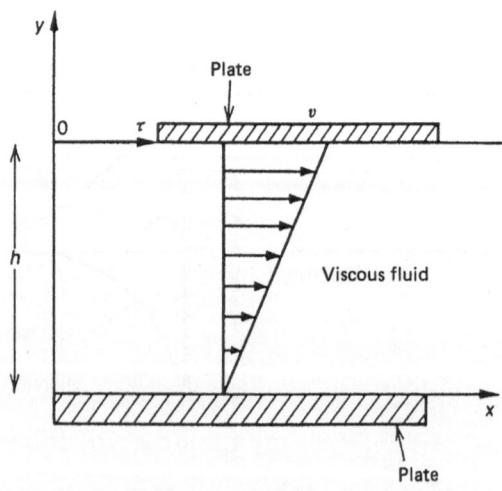

Fig. 3.12 Viscosity of a fluid.

plate is fixed, and when shearing force τ is applied to the upper plate, shifting it parallel to the lower plate at speed v, the fluid in contact with the upper plate flows at the same speed as the plate, but the part of the fluid which is not in contact with the plate moves at a speed lower than v, due to the internal friction within the fluid. The further away from the plate, the higher the deceleration rate. In most fluids, the deceleration rate follows Newton's law of viscosity, which is given below:

$$\tau = \eta \left(\frac{dv}{dy} \right) \tag{3.9}$$

In Eq. (3.9), η is the coefficient which shows the internal friction of the fluid, termed the viscosity. A fluid which obeys this formula is known as a Newtonian fluid. However some fluids show a flow behavior which does not obey the above principle, and the viscosity of such fluids varies according to the shearing force and shearing speed.

Figure 3.13 shows the flow behavior of different types of fluid. Line (a) is a Newtonian fluid, and shearing force and deformation speed are in proportion. In (b) and (c), the relationship between shearing force and deformation speed takes the form of a curve, and such fluids are known as non-Newtonian fluids. Most liquids and gases are Newtonian fluids.

In general, sols are considered to be non-Newtonian fluids. The sol dispersion medium itself is a Newtonian fluid, so the existence of particles and macromolecules in the sol is thought to change a Newtonian fluid into a non-Newtonian fluid. That is to say, particles and macromolecules have an effect on the viscosity behavior of the sol.

If we call the viscosity of the solution η, and the viscosity of the solvent η_0, then $\eta_{rel} = \eta/\eta_0$ is termed the relative viscosity. The rate of increase of the viscosity of the solution relative to the viscosity of the solvent, or $\eta_{sp} = (\eta - \eta_0)/\eta_0 = \eta_{rel} - 1$, is called the specific viscosity, and this shows the increase in the viscosity of the solvent, due to the solute. This effect depends on the solute concentration C, so the rate of increase in viscosity per unit concentration, or $\eta_{red} = \eta_{sp}/C = (\eta - \eta_0)/(\eta_0 C)$ is known as the reduced viscosity, and the value obtained by extrapolating the reduced viscosity to zero is known as the intrinsic viscosity. Generally, there are

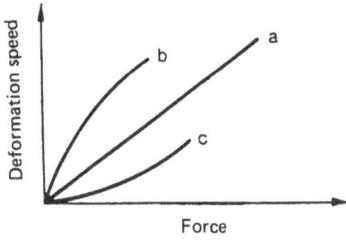

Fig. 3.13 Flow behavior of a fluid. a, Newtonian fluid; b,c, non-Newtonian fluid.

two ways in which the reduced viscosity of a solution – $\eta_{red}=\eta_{sp}/C$ – depends on the concentration. These are shown in Fig. 3.14: (a) is the behavior pattern which usually appears, and when there is strong interaction between the particles, behavior pattern (b) appears. On the other hand, viscosity in relation to spheroidal particle dispersion where solid particles are dispersed in a liquid is expressed by Einstein's viscosity formula: $\eta_{sp}=2.5\phi$. Here, ϕ represents the volume fraction of the particle. This viscosity formula holds true for many particle disperse systems, such as gold colloids and latex. However, at high densities there is a discrepancy between the predicted and the actual values. Many extended formulae have been proposed regarding non-spheroidal particles [10].

3.5.1 Viscosity in Model Disperse Systems

Various types of synthetic resin latex, made with emulsion polymers, are made up of fine monodispersed spheroidal particles with a diameter in the region of 0.1 µm, and they have therefore often been studied as a model for colloid particle disperse systems. Saunders has studied the influence of latex concentration on clay using monodispersed polystyrene latex of 99–871 nm [11]. His results showed that when the latex concentration is 0.25 or less, in volume fraction, a latex disperse system is a Newtonian fluid, and the addition of latex over and above this changes the disperse system into a non-Newtonian fluid. The relationship between reduced viscosity and latex concentration in Newtonian fluids is shown

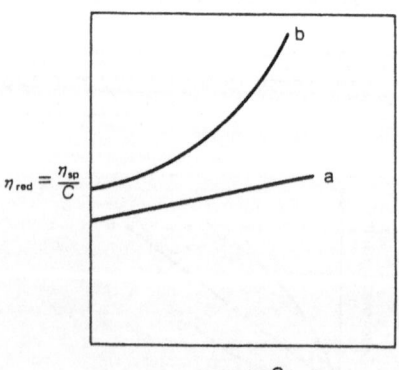

Fig. 3.14 Reduced viscosity $\dfrac{\eta_{sp}}{C}$ and intrinsic velocity $[\eta] = \left[\dfrac{\eta_{sp}}{C}\right]_{C\to 0}$

in Fig. 3.15. It can be seen from this diagram that as the concentration of the latex gets higher, the reduced viscosity increases. Even at the same latex concentration, it can be seen that the smaller the latex particle diameter, the higher the viscosity will be. The relationship between the latex concentration (volume fraction) and viscosity is expressed by Mooney's equation

$$\eta_{rel} = \exp \frac{(\alpha_0 \phi)}{(1 - \kappa \phi)} \tag{3.10}$$

Here, η_{rel} is the relative viscosity, ϕ is the volume fraction, α_0 is the shape factor, here 2.5, the same as Einstein's constant, and κ is the electrostatic attraction constant, which is about 1.35. The diagram also shows the effect of the latex particle diameter on viscosity. It is known that even when the volume fraction is the same, the smaller the latex particle diameter, the higher the viscosity will be. That is to say, the smaller the latex particle diameter, the greater the value for κ, which is the electrostatic attraction constant in Mooney's equation. This means that the smaller the latex particle diameter, the greater the latex particle surface area, and the stronger the effect of electrostatic attraction between the latex particles.

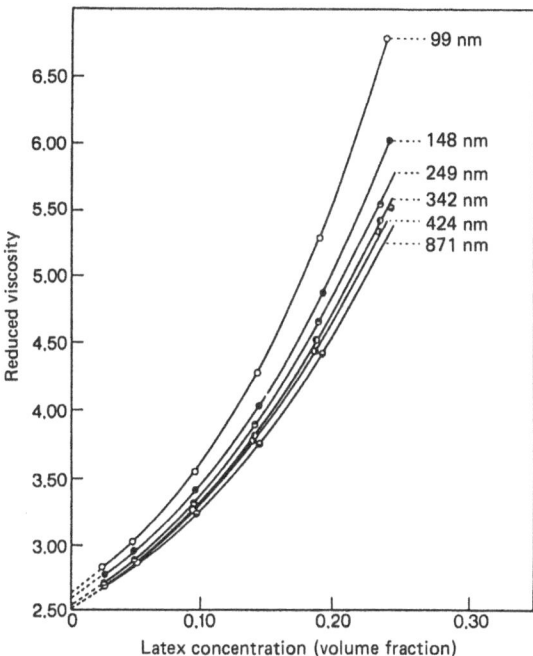

Fig. 3.15 Relationship between latex particle concentration and reduced viscosity.

The influence of particle size distribution on viscosity has also been investigated using polymethyl methacrylate particles of 0.1, 0.6, 1.0 and 4.0 μm mixed in various ratios and monodispersed in Nujol [12]. It was shown that the viscosity corresponds closely to the following equation.

$$\eta_{rel} = \exp \frac{(\alpha_0 \phi_1)}{(1 - \kappa_1 \phi_1)} \cdot \exp \frac{(\alpha_0 \phi_2)}{(1 - \kappa_2 \phi_2)} \cdots \cdots \tag{3.11}$$

The results of this study show that the viscosity of a system consisting of a mixture of particles of different diameters is a hybrid of the Mooney equation for each system.

These relationship equations can provide a good description of the viscosity behavior of disperse systems close to the conceptual model of polyethylene and polymethyl methacrylate latex. However, as we stated in Sect. 3.2.2 and Sect. 3.3, most inorganic particles form an electric double layer when in liquid, and moreover, the particle shape is not regular. Therefore, the viscosity behavior of disperse systems differs considerably from this model. The viscosity behavior of kaolin disperse sytems and magnetic particle disperse systems has been investigated from the point of view of practical applications.

3.5.2 Viscosity in a Kaolin Disperse System

The viscosity–concentration curve [13] for a kaolin disperse system is usually divided into three parts; at low densities a direct relationship has been observed, but the ratio constant is considerably greater than the theoretical value of 2.5 in the Einstein formula. At densities greater than this, the relationship becomes non-linear. In these two areas, thixotropy has been observed, but this is related to the fact that the hydration water on the surface of the particles is removed. When the concentration rises even higher than this, dilatancy is observed. The behavior of kaolin disperse systems in the areas where it shows nonlinearity and dilatancy can be expressed approximately by a Mooney-type equation

$$\frac{1}{\sqrt{\eta}} = 10 \cdot \exp \left\{ \frac{-\alpha_1 V}{2(1 - \kappa_1 V)} \right\} \cdot \exp \left\{ \frac{-\alpha_2 V}{2(1 - \kappa_2 V)} \right\} \tag{3.12}$$

Here, α_1 and α_2 are the constants related to the friction, and κ_1 and κ_2 are the constants related to the packing of the particles. The results obtained for an actual disperse system of ceramic particles are thus different from the results obtained in an experiment using latex. The shape of the particles has a strong effect on viscosity, but this effect is complex and the details are not understood. One factor which is known to exert a strong effect on the viscosity behavior of clays such as kaolin disperse systems in experiments is the effect of coexisting ions. In a disperse system of clay particles, the co-existing ions change the state of the electric double layer, which causes changes in the viscosity of the disperse system.

3.5.3 Effect of the Electrical Double Layer on Viscosity [2]

Ca-clay slip with a solid concentration of about 60% (about 36% in volume concentration) is quite a viscous substance. If NaOH is added to this slip, its fluidity is greatly improved, and if Na_2SiO_3 is used instead of NaOH, the improvement in fluidity is even greater. The changes in the viscosity of Ca-clay due to the addition of NaOH and Na_2SiO_3 can be explained as follows:

$$Ca\text{-}clay + 2\,NaOH \rightarrow clay < \begin{matrix} Na^+ \\ Na^+ \end{matrix} + Ca(OH)_2 \quad \text{(soluble)}$$

$$Ca\text{-}clay + Na_2SiO_3 \rightarrow clay < \begin{matrix} Na^+ \\ Na^+ \end{matrix} + CaSiO_3 \quad \text{(soluble with difficulty)}$$

That is to say, the change in viscosity of Ca-clay by the addition of Na^+ ions, which cause the Ca^+ ions to be removed from the diffusion layer of the clay by being exchanged for Na^+ ions, depends upon the kind of salt being produced. If the salts produced by the exchange are less soluble, the peptization will be more complete. This is because, at a pH of 7 or over, the edges of the clay particles adsorb enough OH^- to give a good electrical charge to the particles, and the multivalent ions in the system completely disappear, thereby making it possible for the double layer to thicken considerably. If the slip is peptized with NaOH, adequate peptization will not occur unless the pH is 12 or over. This is because, at all pH values, clay particles essentially have a negative electrical charge, and positive ions such as Ca^{2+}, Mg^{2+} and H^+ make the double layer thinner, even at a high pH. However, positive ions such as Na^+, K^+ and NH^{4+} increase the thickness of the double layer, leading to low viscosity. Therefore, the multivalent positive ions must be removed from the clay by a method such as precipitation in order to produce a low-viscosity clay.

When water glass (a dense solution of sodium silicate Na_2SiO_3) is used, peptization can be made to occur even at a pH of 8 or under. This is because silicic acid or silicic acid plus SiO_3^{2-} in colloid form are generated by the hydrolysis of sodium silicate as shown in the formulae below

$$Na_2SiO_3 + 2\,HOH \rightarrow 2\,NaOH + H_2SiO_3$$
$$Na_2O \cdot 2\,SiO_2 + 2\,HOH \rightarrow 2\,NaOH + SiO_3^{2-} + H_2SiO_3$$

Accordingly, the negatively-charged colloidal silica particles are adsorbed at comparatively low pH values at the neutral or positive edges of the lamellar clay particles, and even at a pH of 12 or under, they form particle surfaces with a full negative charge.

In an oxide slip, the relationship between zeta potential and viscosity has been thoroughly investigated, and it is known that the higher the zeta potential, the lower the viscosity of the slip will be. Although aluminum chloride is a good peptization agent for alumina, aluminum sulfate is not. Aluminum chloride and aluminum sulfate produce HCl and H_2SO_4, respectively, when hydrolyzed. SO_4^{2-} ions always hinder the peptization of oxides. On the other hand, when $BaCl_2$ is added to such a system, $BaSO_4$ is precipitated, causing the obstructing SO_4^{2+} ions to change places with the Cl^-, and thickening the double layer.

3.5.4 Relationship Between Viscosity and Shearing Stress in a Ceramic Disperse System

Many equations have been proposed to give the relationship between shearing stress and shearing speed and viscosity in fluids. The Ostwald–deWaele equation, however, is the most widely used. This equation can be given by the following formula [9]:

$$\eta = \eta^\circ \left(\frac{\dot{\gamma}}{\dot{\gamma}^\circ}\right)^{n-1} = \eta^\circ \left(\frac{\tau}{\tau^\circ}\right)^{\frac{n-1}{n}} \tag{3.13}$$

In this formula, η° shows the viscosity at shearing stress τ°, and at shearing speed $\dot{\gamma}^\circ$. In a Newtonian fluid, the relationship between shearing stress and shearing speed is a straight line with a gradient of 1. When it is expressed by the above formula, the relationship $\eta=\eta^\circ$ is calculated from $\eta=1$. This shows that viscosity is fixed, regardless of shearing stress and shearing speed. We have already mentioned that colloid solutions and slip show non-Newtonian behavior. Generally, the relationship between shearing stress and shearing speed in slip is as shown in Fig. 3.16. The diagram also shows the apparent viscosity of the slip. When the shearing speed is low, it shows thixotropy, and when the shearing speed is high, it shows Newtonian behavior as well as thixotropy. At even higher shearing speeds, it shows Newtonian behavior and also dilatancy. The Newtonian flow at low shearing speeds is called primary Newtonian flow (I-N), and the Newtonian flow at high shearing speeds is called secondary Newtonian flow (II-N). The existence of a tertiary Newtonian flow has also come to light in recent years [14]. Thixotropy and dilatancy are closely related to the flocculation between the particles. The particle flocculations have two types of structure, as shown in Fig. 3.17. In one structure, the particles are close together and flocculated by a strong force, while in the other, the particles are far

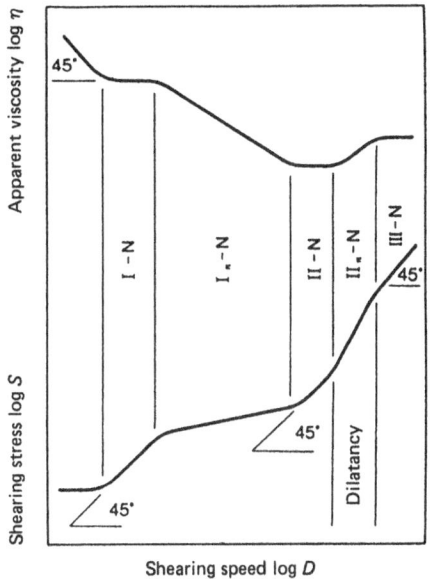

Fig. 3.16 Relationship between shearing speed and shearing stress regarding viscosity.

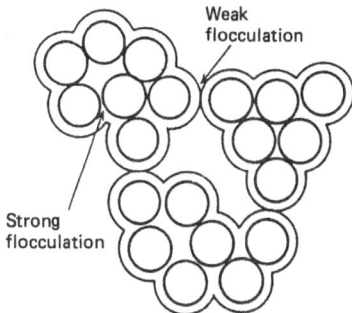

Fig. 3.17 Schematic diagram of flocculation structures.

apart, and the flocculating force is weaker. In normal flocculation, both types of structures coexist. In disperse systems which show thixotropy, weak flocculation plays the principal part, and in disperse systems which show dilatancy, this is closely related to strong flocculation. More generally, the behaviors of thixotropy and dilatancy have been explained in the light of the state of dispersal [15]. Figure 3.18 is a schematic

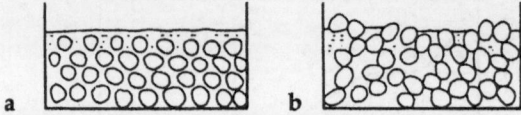

Fig. 3.18 Schematic diagram of dispersal states. **a** Dense structure, and **b** sparse structure.

diagram of the two different particle dispersal states. It is thought that dilatancy is caused by the change in states from a to b, and thixotropy is caused by the change in states from (b) to (a).

3.5.5 Viscosity of Magnetic Fluids

In magnetic fluids, strongly dispersed magnetic powder is subject to magnetic stress in a magnetic field. The movement of the particles is constrained by this magnetic stress, so their rheology is more interesting than that of non-magnetic disperse systems. Figure 3.19 shows the relationship between the viscosity of a magnetic field and the strength of the magnetic field [16]. Curve A shows what happens when the magnetic field is applied parallel to the flow of the fluid, and curve B shows the changes in viscosity when the magnetic field is applied perpendicular to the flow of the fluid. In both cases, the viscosity increases as the strength of the magnetic field increases, as predicted. The viscosity behavior when both magnetic stress and shearing stress are applied to a magnetic fluid is shown in Fig. 3.20 [17]. The vertical axis of the graph shows the viscosity, and the horizontal axis shows the ratio between shearing stress and magnetic stress. This viscosity curve can be divided up into three parts.

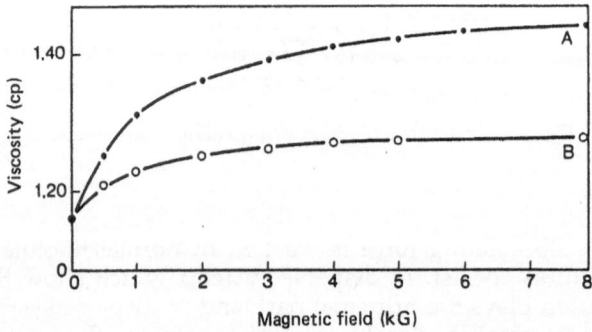

Fig. 3.19 Effect of a magnetic field on the viscosity of a magnetic fluid: A, when the field is applied parallel to the flow, and B, when the field is applied perpendicular to the flow.

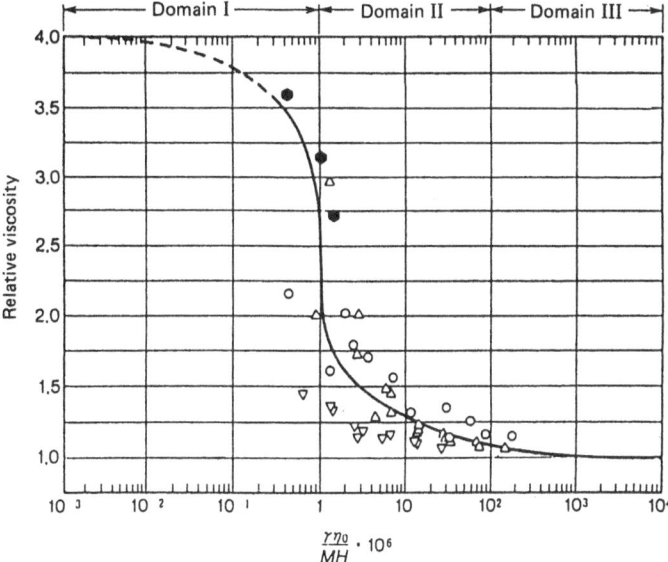

Fig. 3.20 Effect of shearing stress/magnetic stress on the viscosity of a magnetic field. ●, dispersion medium 1 (hydrocarbon); ○, dispersion medium 2 (hydrocarbon); △, dispersion medium 3 (fluorocarbon); ▽, dispersion medium 4 (fluorocarbon).

$$0 < \frac{\gamma \eta_0}{MH} < 10^{-6} \text{ (domain I)}$$

$$10^{-6} < \frac{\gamma \eta_0}{MH} < 10^{-4} \text{ (domain II)}$$

$$10^{-4} < \frac{\gamma \eta_0}{MH} < \infty \quad \text{ (domain III)}$$

In domain III, the viscosity is at its lowest, and is fixed. In domain II, the viscosity depends on the magnetic stress and shearing stress, and in domain I the viscosity is at its highest, and is fixed.

In domain I, the shearing stress is small relative to the magnetic stress, so the movement of the colloid particles in the magnetic fluid is constrained by magnetic stress. As a result, viscosity increases. Conversely, in domain III, where shearing stress is high in relation to magnetic stress, the strong shearing stress causes the particles to move, and the magnetic stress is not strong enough to constrain their movement, so viscosity falls.

In domain II, which lies between the two, the viscosity of the fluid varies widely according to the variations in shearing stress/magnetic stress. When colloids are dispersed in a magnetic fluid, each particle has an outer layer of adsorbed molecules which exert steric hindrance around the particle. The fluid properties of a magnetic fluid are related to the size of the particle, including the adsorbed molecules. Its magnetic properties, on the other hand, are determined solely by the magnetic particles, excluding the adsorbed molecules. Therefore, domain II varies according to the ratio between the size including the adsorbed molecules and the size of the magnetic particles by themselves. For example, in a magnetic fluid with particles where the ratio between the radii is 0.35, domain II will be between 10^{-6} and 10^{-4}, and if the ratio between the radii is 0.45, domain II will be between 10^{-6} and 10^{-8}.

3.6 Gels

If, in a particle disperse system, the dispersion medium is volatilized, or certain kinds of ions are added, the viscosity of the disperse system increases, and finally it loses its fluidity and reaches a gel state. In a gel, the particles which in a sol are dispersed are to some extent agglutinated, but structurally speaking there is no clear dividing line between a sol and a gel.

In recent years, attention has been focused on the production of ceramics by the sol–gel method, which is designed to produce high-performance, highly reliable ceramics using the sol–gel transition in fine particle disperse systems [18]. In this section, we shall look at gels, which are central to this method.

The sol–gel method is a method for producing ceramics, and is carried out as follows: a shape is produced using the fluidity of a sol such as a liquid, after which it is solidified, reaching a gel state without losing the high homogeneity it possessed as a sol. A ceramic is then produced from this gel by sintering. It is also possible to carry out precise control of the fine structure of the ceramics produced, by controlling the composition and diameter of the sol particles. The direct precursor of the ceramic material is the gel, and the following are some of the methods used to convert a sol to a gel: (a) the addition of a gelation agent to the sol; (b) the removal of the solvent from the sol; and (c) the promotion of a chemical reaction in a liquid system. Method (a) is not ideal, because the gelation agent sometimes remains in the ceramic as an impurity. The new sol–gel method is therefore a combination of (b) and (c), in which the starting materials used are organic metal compounds, particularly metal alkoxides. In both of these methods, considerable compression of volume is carried out in the process of converting the sol to a gel, and cracks and fissures tend to occur in the gel during this process, and in industrial applications of the sol–gel method, one of the most important tasks is to develop a technique for producing complete gel bodies with no cracks or fissures. In this section, we shall describe the production, drying, and firing of gels, which are mainly induced from alkoxides. As we said above, there is no great difference between the structure of a sol and that of a gel, and the structure of a gel is an extension of that of a sol. Therefore, in structural terms, sols and gels overlap to some extent, and we shall describe both.

3.6.1 Producing Gels

Metal alkoxides can be described by the general formula $M(OR)_n$. In this formula, M is the metallic element with value n, and R is the alkyl group. Metal alkoxides can easily be made to hydrolyze by causing them to react with water, but the hydrolysis reaction is in general complicated by the interaction between the substance generated by hydrolysis and the metal alkoxide, or between the substances generated by hydrolysis. That is to say, the hydrolysis system includes the following two basic reactions: (1) the hydrolysis reaction and (2) the polycondensation reaction.

$$-M-OR + HOH \rightarrow -M-OH + ROH \text{ (hydrolysis reaction)} \tag{1}$$

$$-M-OH + RO-M- \rightarrow -M-O-M- + ROH \text{ (polycondensation reaction)} \tag{2}$$

Many metal alkoxides react very easily with water, so the reaction takes place rapidly in such a way that the alkoxy group of the system uses up all the water present. However, in alkoxides such as silicon alkoxide, which are poorly reactive with water, the reaction with water takes place very slowly, and the structure of the substance generated by hydrolysis includes quite a large amount of the alkoxide group OR. The question of whether any alkoxy group or OH group remains in the structure of a substance generated by hydrolysis thus depends primarily on the chemical properties of the metal element. In alkoxides where the substance generated by hydrolysis is an oxide, there is little tendency for OH groups, OR groups and so on to remain in the structure, and if they do, they are present only on the surface of the particles. On the other hand, when the substance generated by hydrolysis is a hydrate, the amount of OH, OR and similar groups remaining in the structure can be considerable. This means that the inclusion of groups such as OH and OR in the –M–O–M– skeleton is controlled by the chemical properties of the –M–O–M– skeleton. On the other hand, the structure of the –M–O–M– skeleton formed is strongly influenced by the chemicals generated by the hydrolysis reaction. For example, it is thought that the hydrolysis of $Si(OR)_4$ produces four basic kinds of chemicals: $Si(OR)_3(OH)$, $Si(OR)_2(OH)_2$, $Si(OR)(OH)_3$ and $Si(OH)_4$. Several different types of –Si–O–Si– skeletons of different sizes are produced by the reaction of these chemicals, and these continue the polycondensation reaction, making it even more complicated. The basic chemicals produced at the start of the hydrolysis reaction, and the type and size of the skeleton structure which follows are influenced by the reaction probability of the kinds of reaction which are present in the reaction system. The basic factors which determine the dynamics of the reaction are therefore:

1. The water/alkoxide ratio
2. The concentration of water and alkoxide in the solution

Figure 3.21 shows an experimental calculation of the way in which the water/alkoxide ratio influences the oxide content of the sol particles produced by hydrolysis. These data alone do not shed direct light on the shape and size of the sol particles present in the sol. However, Yoldas has pointed out that the shape and size of the sol particles can be indirectly inferred by considering the following model structure, using silica sol as an example. Let us assume that the sol particles have radius r, are spheroidal, and contain n Si atoms, and that only the Si atoms on the particle surface are linked to the OH radials. The volume of the sol particles is proportional to the number of Si atoms, and the number of surface OH radicals is proportional to the surface area of the particle, so

$$\text{Number of OH radicals} \simeq 4\pi r^2 = 4\pi \left(\frac{3n}{4\pi}\right)^{\frac{2}{3}} = (36\pi n^2)^{\frac{1}{3}}$$

Also, since two oxygen atoms are linked to one Si atom, the number of bridging oxygen atoms in the particle can be calculated by subtracting half the number of surface OH groups from twice the number of Si atoms, and the formula is as shown below:

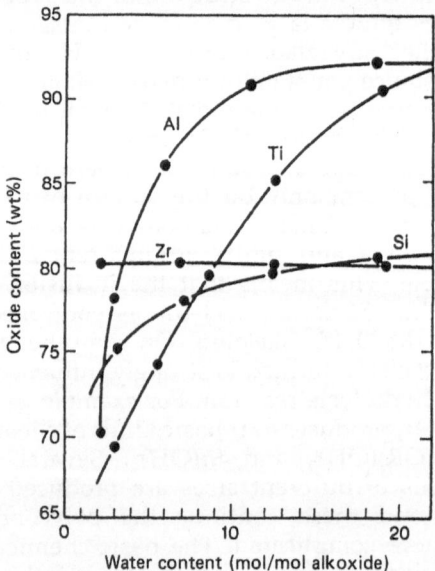

Fig. 3.21 Oxide content of product obtained by hydrolysis of various alkoxides at different water/alkoxide ratios.

$$\text{Number of bridging oxygen atoms} = 2n - \frac{(36\pi n^2)^{\frac{1}{3}}}{2} = 2n - \left(\frac{9}{2}\pi n^2\right)^{\frac{1}{3}}$$

The chemical formula for the sol particles can therefore be expressed as follows:

$$Si_nO_{2n - (\frac{9}{2}\pi n^2)^{\frac{1}{3}}}(OH)_{(36\pi n^2)^{\frac{1}{3}}}$$

These particles are pyrolyzed by heating, as shown below:

$$Si_nO_{2n - (\frac{9}{2}\pi n^2)^{\frac{1}{3}}}(OH)_{(36\pi n^2)^{\frac{1}{3}}} \rightarrow nSiO_2 + \left(\frac{9}{2}\pi n^2\right)^{\frac{1}{3}} H_2O$$

The oxide content of the sol particles is related in the following way to the number of Si atoms n contained in a sol particle, i.e. to the size n of the sol particle

$$\text{oxide content of sol particle} = \frac{nSiO_2}{nSiO_2 + \left(\frac{9}{2}\pi n^2\right)^{\frac{1}{3}} H_2O}$$

$$= \frac{60n}{60n + \left(\frac{9}{2}\pi n^2\right)^{\frac{1}{3}}} = \frac{20}{20 + 3\left(\frac{36\pi}{n}\right)^{\frac{1}{3}}}$$

Figure 3.22 shows hypothetical models of the three types of skeleton: one-dimensional (linear), two-dimensional (planiform) and three-dimensional (spheroidal) and shows their oxide content as related to the size of the sol particles. This curve is very similar to the experimental curve shown in Fig. 3.21 for the oxide content of the product of hydrolysis measured for different water alkoxide ratios, for different alkoxides: it shows that the concentration of water and alkoxide has a strong influence on the dynamics of the reaction. However, we must not overstate the resemblance between the curves based on skeleton models like this, and curves arrived at by experimentation. The decrease in oxide content accompanying the decrease in the size of the particles in the model is the natural result of the fact that OH radicals are present only on the particle surface. On the other hand, the decrease in the oxide content accompanying the decrease in the water/alkoxide ratio in the results of the experiment is thought to be more strongly influenced by the OR radicals remaining inside the particles due to incomplete hydrolysis rather than by the decrease in the size of the particles. This is also shown by the fact that a gel produced from a sol which has been hydrolyzed with a small amount of water is more prone to have traces of carbon left over from pyrolysis, and cracks and fissures, than is a gel which has been produced from a sol hydrolyzed with a large amount of water.

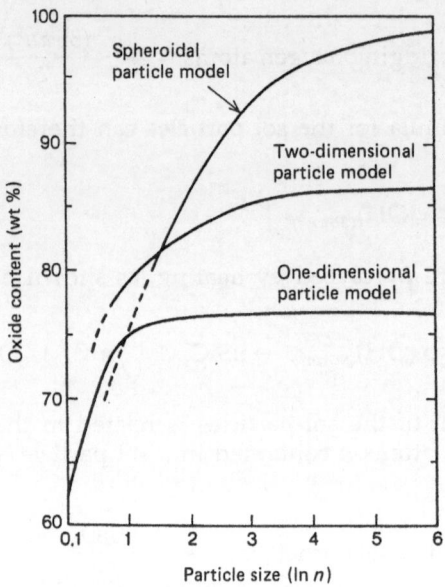

Fig. 3.22 Relationship between size and oxide content of gel particles obtained by hydrolysis of an alkoxide.

The shape and size of sol particles generated by the hydrolysis of a metal alkoxide, and accordingly the structure of the gel, are also affected by the pH of the system. The extent to which they are affected varies widely depending on the type of metal element. For example, the hydrolysis reaction of a Si alkoxide is strongly affected by pH, but the hydrolysis reaction of Zr alkoxide is hardly affected at all. The case of Al alkoxide is intermediate between these two. The cause of this difference in the product of hydrolysis is thought to be the fact that the reactions in the system differ according to the pH. For example, in the hydrolysis of $Si(OC_2H_5)_4$, at low pH, the electrophilic reaction involving the oxygen in the alkoxide is the controlling factor, while at medium to high pH values, the nucleophilic reaction is the controlling factor. As we stated earlier, the product of hydrolysis is influenced by the water/alkoxide ratio. It is also known that when Si alkoxide is used and the water/alkoxide ratio is 4, the product of hydrolysis has a chain structure. The flocculation state of this chain structure also varies depending on the pH of the liquid. Small-angle X-ray scattering experiments were carried out, keeping the water/alkoxide ratio the same and varying the pH. These experiments demonstrated that, as shown in Fig. 3.23, when the pH is 1, the chain structure is mainly simple, with little branching, and most of the individual chains are separate. However, when the solution has a pH of 7, there is a good deal of branching, and the chains, which branch in an intricate manner, intertwine and form clusters. This structure generally

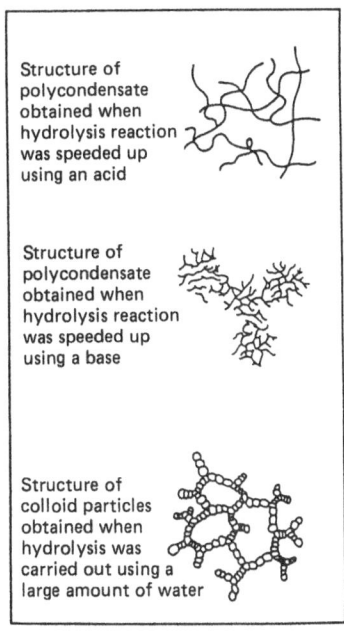

Structure of polycondensate obtained when hydrolysis reaction was speeded up using an acid

Structure of polycondensate obtained when hydrolysis reaction was speeded up using a base

Structure of colloid particles obtained when hydrolysis was carried out using a large amount of water

Fig. 3.23 Structure of colloid particles and polycondensate obtained by hydrolysis of Si alkoxide.

occurs when there is a small amount of water and the pH is high: under these conditions, the viscosity of the sol is high, and with the passage of time, the products of hydrolysis intertwine and form bridges, and the sol becomes a gel. A gel formed in this way does not thereafter undergo any further structural changes worthy of note. Conversely, when hydrolysis is carried out using a large amount of water, under normal circumstances the well-known colloidal silica solution is obtained. As shown in the diagram, nearly all those silica particles with a structure closely resembling an oxide skeleton then form bridged three-dimensional networks. This is because, in a system which includes a large amount of water, a substantial proportion of the particles are dissolved and precipitated, the particle skeleton structure becomes more highly concentrated, and more closely approaches an oxide structure. However, gelation by removing the solvent from a sol with a low solid concentration produces particles with a lower bulk density than those with a chain structure, because of the large pores formed in the skeleton structure. As we stated before, the hydrolysis of Si alkoxide is influenced by many factors and is extremely complicated, but a summary of the results obtained to date concerning silica gel production is given in Fig. 3.24. The main reason why the hydrolysis of Si alkoxide is so complicated is that it is slow, so the ratio of OH and OR groups included in the structure

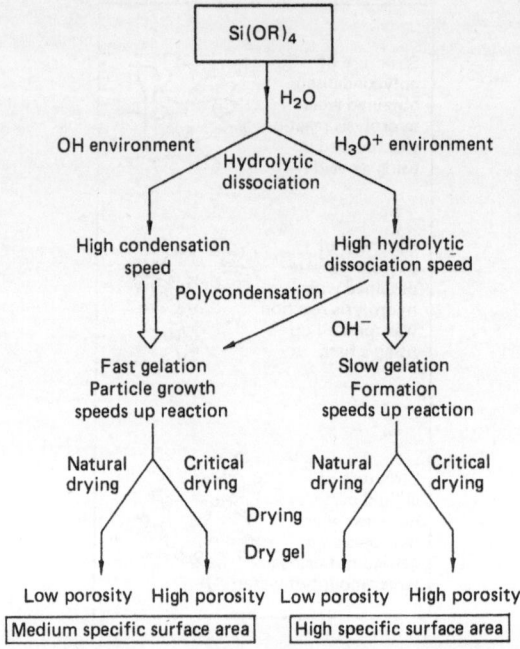

Fig. 3.24 Two basic methods for obtaining dry gel from Si alkoxide using the sol–gel method.

of the product of hydrolysis can vary considerably. On the other hand, many ordinary metal alkoxides hydrolyze very rapidly, the hydrolysis reaction being completed almost instantaneously, and therefore, even if the factors involved in the system are controlled, it is impossible to control the reaction effectively. For this reason, there is less variation in the structure of these sol particles than there is when Si alkoxide is used, and there is also much less variation in the skeleton structure of the gels formed from such sols.

3.6.2 Drying Gels

In order to apply the sol–gel method to practical ceramics production, it is necessary to obtain a molded body of dry gel which is free of cracks and fissures, as a precursor before firing. The technical difficulty or otherwise of obtaining a fault-free body of dry gel depends on the type of gel, the shape and size of the gel particles, the gel structure, and the desired macroscopic shape of the dry gel. As a sol or a gel becomes a dry gel, the viscosity of the system increases, and passes the point where rapid solidification occurs. Many gels crack at this point. A simpler method of obtaining a dry gel is to dry the gel naturally by exposing it to the atmosphere, but to obtain a completely dry gel by this method, it must be dried extremely slowly directly before solidification begins. This is because, in proportion to the flow of the solvent on the surface, and the depth of the sample, surface stress is generated in inverse proportion to the dispersal coefficient of the solvent on the surface. That is to say, when the evaporation of the solvent from the surface is accelerated, a large surface stress is generated and the sample cracks. There are no quantitative data regarding this phenomenon, but it is probably desirable to have the gel pass this solidification point at a drying speed no greater than 10% to 20% per day. However, the difficulty of obtaining a perfect body of dry gel increases rapidly in relation to the desired size of the gel body. It is difficult to generalize about quantitative drying methods, but the actual drying curves for SiO_2 alkogel (gel produced using alcohol as the dispersion medium) shed some light in the form of experimental values (Fig. 3.25). In order to prevent cracking, gel drying is usually carried out in a container with a small opening. However, when the process is carried out in the state shown by curve A in the diagram, the solvent is evaporated at a fairly high speed until the gelation begins, but evaporation slows down when gelation starts. After the gelation point (indicated by the arrow), the evaporation speed more or less levels off, and 400 hours are needed before the gel weight loss becomes too small to be measured. Conversely, curve B shows the situation when accelerated drying is carried out by controlled removal of the solvent vapor inside the container. In this situation, it is important to carry out accelerated drying after the gelation point (indicated by the arrow) in order to prevent the gel from cracking. In this experiment, a perfect 100% crack-free dry gel was obtained as long as the weight loss did not exceed 0.8g/h. The time needed for drying was shortened to 200 hours. In both cases, the proportion of solvent lost before the weight loss became too small to be measured was about 60%.

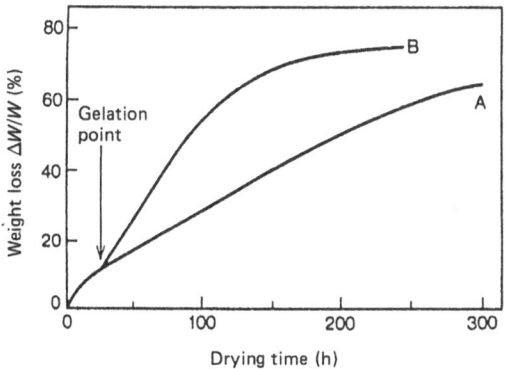

Fig. 3.25 Drying curve for SiO_2 alkogel.

Besides the surface stress generated as the solvent evaporates from the surface, another important stress which causes cracking is the capillary pressure within the gel. The level of the capillary pressure is proportional to the specific surface energy at the liquid–gas interface, and inversely proportional to the diameter of the capillaries. That is to say, the smaller the diameter of the capillaries, the greater the capillary pressure. Figure 3.26 shows the capillary pressure acting on the water-filled capillaries, and the shearing stress it generates. The capillaries formed in a dry gel are very small, many of them being less than 10 nm. The capillary pressure therefore reaches several hundred atmospheres. When the capillaries are equal in size and dispersed evenly throughout the gel, their presence does not cause cracks in the gel. Cracking occurs when a large and a small capillary are formed close together and a large pressure differential acts between the two. The distribution of pore diameter throughout the gel is a function of the gelation process, and normal slow drying gives a tighter pore diameter distribution and a smaller pore diameter than fast drying. This is another reason why slow drying can prevent cracking. In the production of dry gel by natural drying, the drying speed is kept very low in order to prevent the generation of surface stress due to the evaporation of the solvent, and also to prevent the generation of capillary pressure due to uneven distribution of pores in the gel. In order to produce perfect gels at high drying speeds, various steps are therefore taken.

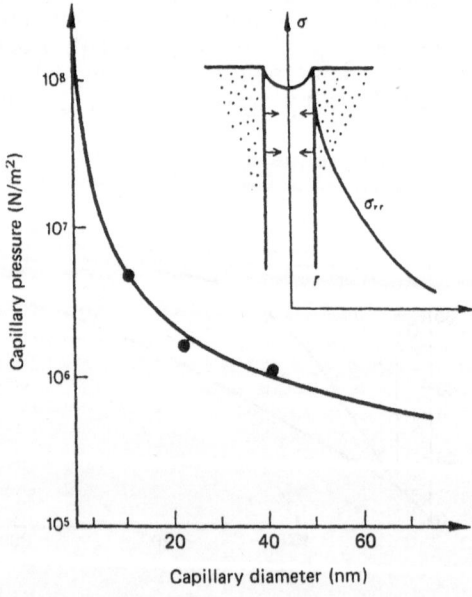

Fig. 3.26 Capillary pressure and state of stresses around a capillary wall.

In the critical point drying method, a fairly large volume of gel can be dried at a comparatively high speed. Figure 3.27 shows the curve for the equilibrium of the solvent between the liquid phase and the gas phase. Unlike the normal drying operation, in which the transformation of the solvent from liquid to gas is used, the critical point drying method follows the route a→b→d→e or a→d→e. That is to say, without cutting across the solvent's gas–liquid equilibrium curve, the solvent's critical point c is avoided by a detour. The liquid is therefore vaporized continuously, and removed from the gel. This operation can be carried out experimentally using an autoclave. The gel and the required amount of solvent are placed in the autoclave, which is then sealed. Heating is then carried out, and the system then moves to point d. When the solvent has been removed from the system over a period of several hours, a dry inert gas such as argon is passed over the gel, completely removing any traces of solvent remaining in the gel. After this, the temperature of the system is lowered, and the gel is removed from the autoclave when it has reached room temperature. By controlling the heating speed and the amount of solvent, it is possible to obtain a large body of dry gel with a diameter of several centimeters and a depth of several tens of centimeters, with several hours drying without fail. With the critical point drying method, even if the gel includes a fairly large amount of solvent, as long as the gel particles have a three-dimensional skeleton structure, a gel can be obtained which is free of cracks and fissures. However, this method is not suitable for obtaining high density dry gel bodies. This is because, unlike the natural gel drying method (see the curve in Fig. 3.25), where the density of the gel is increased considerably by capillary pressure after gelation also, in the critical point drying method, there is no separate force to make the gel contract. This is the drawback of the critical point drying method, but experiments have

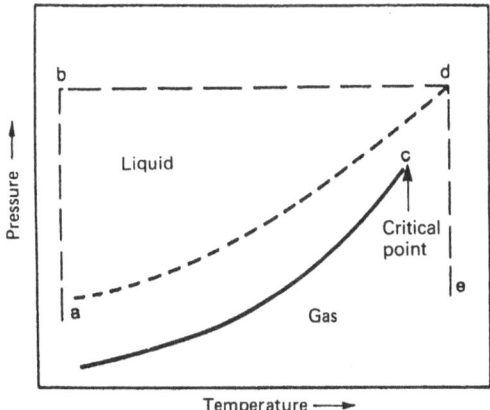

Fig. 3.27 Temperature and pressure route and solvent gas–liquid equilibrium curve for critical point drying.

shown that it is not a particular hindrance when producing high density sintered bodies.

3.6.3 Firing Gels

The firing of gels obtained by the sol–gel method is affected by the skeleton structure of the gel. Many researchers have studied the way the gel skeleton structure affects the increase of the density of the gel. Figure 3.28 shows the heating shrinkage of two gels with the same constitution ($71SiO_2$, $18B_2O_3$, $7Al_2O_3$, $4BaO$) synthesized at different pH levels. Because these two gels have the same composition, the difference in their heating shrinkage curves is probably a direct reflection of the difference in their skeleton structures. That is to say, gel 1 was synthesized at a low pH and a small water/alkoxide ratio, so it has a weak, coarse skeleton structure and shows large shrinkage at low temperatures. Gel 2, on the other hand, was synthesized at a high pH and a large water/alkoxide ratio, so its skeleton structure is closer to an anhydrous oxide, and at low temperatures it does not get much denser. Many factors contribute to the heating shrinkage of the gel, other than the skeleton structure, including the gel composition, pore diameter, and heating speed. Figure 3.29 shows heating shrinkage curves measured for a range of gels with the same gelation conditions but different compositions. The immediately striking point about this graph is that the contraction curve of the gel can be divided into two: a low-temperature domain in which the

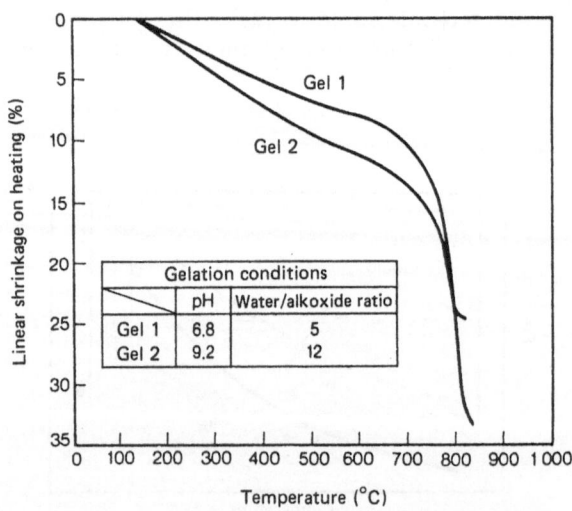

Fig. 3.28 Heating shrinkage curve for dry alumina borosilicate glass with different skeleton structures.

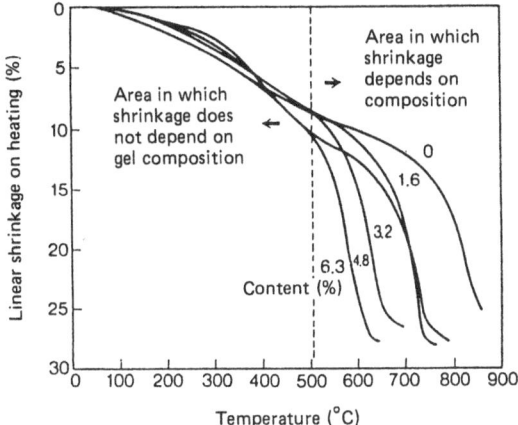

Fig. 3.29 Linear shrinkage on heating curve for borosilicate gels with varying Na₂O content.

contraction has almost no relation to the gel composition, and a high-temperature domain, in which the contraction is strongly dependent on the gel composition. In glass-composition gels like these, at high temperatures, contraction is controlled mainly by the viscosity sintering mechanism, and at low temperatures, mainly by the increase in the density of the skeleton and the decrease in porosity. This is because viscosity sintering is strongly dependent on gel composition, but hardly dependent at all on the heat-induced increase in the density of the skeleton or on the decrease in porosity. The gel skeleton is very imperfect and includes a large number of fine pores, but if it is possible to make the skeleton denser and reduce the porosity at the same time as producing viscous flow, then it is possible that the sintering temperature of the system can be lowered. In ordinary sintering, the increase in the density of the skeleton and the decrease in porosity is dependent on the mass transport, so in dynamic terms, the increase in the heating speed raises the sintering temperature. However, it seems that with gel sintering the increase in density of the skeleton and the decrease in porosity is more dependent upon the starting temperature than a high temperature, thus allowing these processes to take place at the same time as the viscosity sintering mechanism, thus making it possible to lower the sintering temperature. It is reportedly possible to lower the sintering temperature considerably by using a fast heating speed of 10 to 15°C/min.

References

1. R.K. Iler: The Chemistry of Silica, John Wiley & Sons (1979)
2. Y. Ozaki et al.: Ceramic Processing, pp. 106–114, Kihoudou Shuppan (1984) (in Japanese)
3. R.E. Hostetler, J.W. Swanson: J. Polym. Sci., 12, pp. 29–43 (1974)
4. S.S. Papell: US Patent No. 3215572 (1965)
5. R.E. Rosensweig, J.W. Nestor, R.M. Timmins: A.I.Ch.E., Chem. E. Symposium, 5, p. 104 (1965)
6. J. Shimoiizaka: Funtai oyobi Funmatsu Yakin, 25, 6, (1978) (in Japanese)
7. R.S. Sapieszko, E. Matijevic: J. Colloid Interface Sci., 74, 2, pp. 405–422 (1980)
8. L.L. Hench, D.R. Ulrich (eds.): Ultrastructure Processing of Ceramics, Glass and Composites, Wiley-Interscience (1984)
9. K. Itou (trans.), J.M. McKelvey: Koubunshi Kakou Kougaku, Gendai Kougakusha (1962) (in Japanese)
10. T. Nakagawa: Rheology, Iwanami Zensho (1960) (in Japanese)
11. F.L. Saunders: J. Colloid Sci., 16, pp. 13–22 (1961)
12. C. Parkinson, S. Matsumoto, P. Sherman: J. Colloid Interface Sci., 33, 1 (1970)
13. K.M. Beazley: J. Colloid Interface Sci., 41, 1 (1972)
14. K. Umeya: Nihon Gakujutsu Shinkokai No. 136 Committee Futeikei Taikabutsu Shikjou Gijutsu Kyougikai Kenkyuu Shiryou (1978) (in Japanese)
15. S. Oka: Rheology Nyuumon, Kougyou Chousakai (1976) (in Japanese)
16. J.P. McTague: J. Chem. Phys., 51, 1 (1969)
17. R.E. Rosensweig: J. Colloid Interface Sci., 29, 4 (1969)
18. Y. Ozaki: FC Report, 3, 11, pp. 5–12, Fine Ceramics Kyoukai (1985) (in Japanese)

Additional References

19. B.E. Voldas: "Modification of oxides by polymerization process", Proceedings of the International Conference "Ultrastructure Processing of Ceramics, Glasses and Composites", Gainesville, 13–17 Feb. 1983, p. 60
20. C.J. Brinker, G.W. Scherer: "Relationships between the sol-to-gel and gel-to-glass conversions", loc. cit., p. 43
21. C.J. Brinker, W.D.Drotning, G.W. Scherer: "A comparison between the densification kinetics of colloidal and polymeric silica gels", Better Ceramics through Chemistry (ed. C.J. Brinker, D.E. Clark, D.R. Ulrich), p. 25, North-Holland (1984)
22. M. Prassas, L.L. Hench: "Physical-chemical factors in sol–gel processing", Proceedings of the International Conference, "Ultrastructure Processing of Ceramics, Glasses and Composites", Gainesville, 13–17 Feb. 1983, p. 100
23. J. Zarzycki: "Monolithic xero and aerogels for gel–glass processes", loc. cit., p. 27
24. D. Segal: Chemical Synthesis of Advanced Ceramic Materials, Cambridge University Press (1989)
25. R.M. Laine (ed.): Transformation of Organometallics into Common and Exotic Materials: Design and Activation, Martinus Nijhoff (1988)
26. A. Roy, R. Roy: Mat. Res. Bull., 19, pp. 169–177 (1984)
27. A. Makishima: J. Am. Ceram. Soc., 69, (1986) pp. C72–74
28. W.A. Yarbrough, T.R. Gururaja, L.E. Cross: Am. Ceram. Soc. Bull., 66, pp. 692–698 (1987)
29. W.J. Dalzell, D.E. Clark: Ceram. Eng. Sci. Proc., 7, pp. 1014–1026 (1986)

Chapter 4

Physical Manufacturing Processes

Physical methods of producing superfine particles from a range of different materials from metals and ceramics through to organic compounds are currently being devised with a view to the potential applications of the final product.

Superfine particles with physical properties determined principally by their surface or microscopic structures call for methods of generation in keeping with such potential applications. It is thus equally important to find ways of handling superfine particles in keeping with the generation process.

In this chapter we shall discuss various methods of generating superfine particles and shall also touch on certain aspects of the latest particle handling techniques. The text will be clarified as we go along by means of numerous examples of the processes and techniques involved.

4.1 Introduction

The most appropriate physical method for the production of superfine particles is normally determined in accordance with the required size of the final product. There are currently two main methods in use. The first involves the milling of material in its bulk solid form and the second involves the build-up of superfine particles through the aggregation of individual atoms and molecules. The milling method is basically an extension of the use of conventional methods down to micrometer sizes while the build-up method (typical physical methods being evaporation and condensation) entails growth (by particle agglutination, for example) from extremely fine levels of around 2–3 nm, just a little above the cluster[1] level. The two methods thus work in opposite directions in terms of particle size with the milling method working down from sizes in the order of several micrometers and the build-up method working up from sizes in the region of just a few nanometers.

There are currently a large number of different methods being tested for the production of superfine particles including a particularly large number of physical methods (including mechanical methods). Early interest in the production and use of superfine particles was essentially restricted to those involved in physical and chemical research (researchers were primarily interested in the morphological and crystalline characteristics of superfine particles along with their physical properties) but this has extended of late to include researchers and technicians representing a broad cross-section of science and industry from fields such as electrics, electronics, machinery, ceramics and macromolecular chemistry through to biotechnology and medicine.

A considerable number of physical methods are now being tested for the production of superfine particles from a wide variety of different materials, ranging from the more traditional inorganic chemical compounds such as metals and their alloys and ceramics through more recently to organic chemical compounds, macromolecules and solidified gases [1,2]. Most of these, however, are still at the laboratory research stage. The current situation is thus one where research and testing is being carried out over a wide variety of fields into the physical characteristics and uses for superfine particles in powder[2] form.

Researchers are interested in the possibility that materials may exhibit different characteristics in their superfine state from those which they exhibit in their bulk solid state and technicians are interested in identifying such characteristic differences in superfine powders and in finding suitable applications for such differences.

It is thus likely that when a particular application is eventually identified for a superfine particle (superfine powder) then there will be a further examination of its characteristics, and production methods will be considered again in the light of the specific characteristics of the powder concerned so that the most appropriate method may be found for its production. At this juncture (with industrial applications and production methods not yet fully established), therefore, it is probably still too early to be discussing superfine particles in terms of their merits and demerits from the standpoint of industrial production. More appropriate perhaps would be a consideration of ways of improving

---

[1] "Clusters" are ultra-superfine particles containing small, countable numbers of atoms and have become a focus of interest for physical and chemical researchers.

[2] Following the development of applications for superfine particles it has become appropriate to use the name "superfine powder" to describe superfine particles when they are handled in bulk.

current production methods and handling techniques in readiness for larger scale industrial production at a later date.

For example, the technique of generating superfine particles by evaporation in a gas phase was used frequently during the course of the Hayashi Superfine Particle Project [3] (under the overall direction of C. Hayashi, Chairman of Nippon Shinku Gijutsu (KK)), one of the creative science and technology promotion projects that have helped stimulate enormous interest in superfine particles over a broad spectrum of industry in recent times. The reason for the predominant use of this method is that it was already known to possess a number of advantages in that superfine particles produced by evaporation in gas are (a) clean, (b) fairly uniform in size with a sharp distribution, and (c) easily controlled in terms of size.

The kinds of superfine particles used in research and in the search for new applications need to be generated in such a way as to leave them with fixed characteristics, and gas evaporation was felt to be the best way to achieve this end. The gas evaporation method was formerly used to make particulates of metals and their compounds but the Hayashi Project also made use of this same production method with ceramics, complex chemical compounds and organic compounds.

As explained above, Hayashi recognizes the appropriateness of the gas evaporation method for the production of superfine particles for research and development and the identification of new applications and points out that this may well be one of the gaseous phase methods (physical methods) which is eventually developed for use in the industrial production of superfine particles with clearly specified industrial applications [4]. Kazuo Akashi (an emeritus professor at Tokyo University) has pointed out five primary conditions for the production of superfine particles: (a) surface purity, (b) particles must be capable of control in terms of size, (c) they must be collected easily, (d) they must be stable with good conservation characteristics, and (e) they must be easy to produce in industrially usable quantities [5]. Physical production methods are able to fulfil most of these conditions. A number of industrial processes are discussed in more detail in Sect. 4.4.

4.2 Milling Method

The method whereby materials are reduced to particle size from their bulk solid form has long been used and is indeed still used by people involved in the production of powders. If we consider the range of possible sizes for particles used in powders as anything from 1 mm down to 10 μm then particles for fine powders (these should be called fine particles within the terms of this book's title) will fall within a range from 10 μm to 0.1 μm, representing a size graduation of two decimal places for each category. Bulk materials are normally comminuted by mechanical means although there are a great many different mechanisms within this broad general category.

These various mechanisms are collectively known as milling and involve the creation of fine particles from the surfaces of small particles as they are rubbed against each other in their powder form. There are two main ways in which milling is normally achieved. One involves the introduction of powder into a medium such as an agitator of some kind and the other involves the use of high speed jets of air to apply extreme pressure and friction at the same time. Nevertheless as these are both mechanical methods of milling or otherwise comminuting particulates there tends to be lower limit on achievable particle size in the region of 3 μm. This is sometimes referred to as the "3 μm wall".

Even in powders with an average particle size of 3 μm there are nevertheless likely to be particles with diameters of 1 μm or less and for this reason screening is also frequently employed to ensure a more uniform particle size. The composition of a particulate is also a very important element in the milling process (for processing, a particulate should ideally be composed of brittle materials with good heat resisting characteristics). A good example would be the sort of metal which forms a brittle hydride on absorption of hydrogen gas.

However, comminution by mechanical means involves physical contact with parts of the milling machinery, resulting in the admixture of impurities with the particulate. Related problems, including, for example, the presence of pollutants in the processing gas atmosphere, suggest that the best results will be obtained by comminuting not all materials but rather only such materials as seem most appropropriate in terms of their physical characteristics and cost effectiveness (economic characteristics) along with their proposed applications.

4.3 Build-up Method

4.3.1 Outline

This method involves the generation of superfine particles by agglomeration (build-up) of atoms or molecules of the individual elements concerned. The method itself depends basically on the heating and evaporation of selected raw materials to reduce them to their atomic or molecular form. This is followed by multiple condensation in order to generate the required superfine particles. As illustrated in Table 2.1, the number of atoms or molecules required to generate a single superfine particle is currently thought to be in the region of 4×10^3 atoms for what is thought of as the lower limit for superfine particles of 5 nm diameter, and somewhere around 3×10^7 atoms for particles at the upper end of the superfine particle range with diameters of 0.1 μm (100 nm).

Since atoms tend to flocculate in this way during evaporation and condensation it is possible to generate particles with diameters of, for example, 1 μm but the method is not in itself efficient. Basically this method is best suited to the production of particles falling within the size range encompassed by the broad consensus definition of a superfine particle.

Figure 4.1 shows a photograph of clusters of atoms (evaporated) in a superfine particle produced by the build-up method. The photograph itself was taken by Sumio Iijima (at the time a member of the Research Development Corporation of Japan of the Hayashi Superfine Particle Project and currently with Nippon Denki (KK)) using an electron microscope [6]. Iijima's research involved the observation at atomic level of the surfaces and interiors of superfine particles through an electron microscope. The superfine particles used in this research were produced by the build-up method (to be more precise, by the gas evaporation method).

Fig. 4.1 Typical electron microscope photograph [6] showing the agglomerated atomic structure of a superfine particle produced by the build-up method (photograph by Sumio Iijima).

Over a number of years researchers have used not only gas evaporation but also a variety of new methods and mechanisms in order to generate (produce) or extend the range of superfine particles. Some of the principal methods used will now be explained along with some details of the superfine particles produced.

The first series of experiments dealing specifically with superfine particles were probably those performed almost half a century ago, around the time of the Second World War, by Ryoji Uyeda (emeritus professor of Nagoya University and adviser to the Hayashi Superfine Particle Project) during the course of which he produced Zn medium (superfine particles). In order to produce this medium, Zn was evaporated in a vacuum evaporator from which roughly half the air had been drawn off. This resulted in the emission of smoke and the deposition of a black Zn soot on the walls of the evaporator. The evaporating source in this experiment was a W wire which reacted with the oxygen in the atmosphere under heat and was found, in electron diffraction tests, to produce, amongst other things, WO_3. In order to prevent this kind of oxidation the Zn was subsequently evaporated in an alternative atmosphere of N_2 gas which resulted in the generation of a pure Zn medium [7].

This Zn medium (superfine Zn particles) was produced in order to make use of the special properties of infrared rays. This research, which resulted, amongst other things, in the generation of superfine particles through the evaporation of metals including Ni and Fe in inert gases such as He and Ar, is thought to have been the first experimental use of the so-called gas evaporation method.

In the early 1960s, some 20 years or so after these early experiments which resulted in the generation of superfine particles of Zn, a fresh round of research into superfine particles was initiated by a group of physicists again led by Uyeda. Two factors which gave strong impetus to these later research efforts were the contemporary scientific interest in the Kubo effect [8] along with growing use of the new and powerful electron microscope which was also being developed at the time. The publication of photographs taken using an electron microscope enabled Uyeda and Kazuo Kimoto (professor of the Faculty of Arts of Aichi Gakuin University) to demonstrate the variety of platelike or multi-faceted crystals of superfine particles [9,10] to a wider audience, thereby arousing their interest and support for this line of research.

A further boost to this research effort came with the publication of two studies: (a) the evaluation of superfine particles as magnetic recording materials by Akira Tasaki (professor of Tsukuba University) and (b) the use by Nobuhiko Wada (a lecturer at Nagoya University) of a plasma heating technique for the generation of superfine particles. The generation by Tasaki under laboratory conditions of (magnetically) chained superfine magnetic metal particles served to substantiate the existence of high performance magnetic properties in such particles [11]. Superfine particles had previously been generated in milligram quantities but Wada's use of plasma heating under experimental conditions, which was the first instance of the use of continuous heating (for periods of about 1 hour), generated the target materials in the order of several tens of grams [12] thereby serving to stimulate interest in the potential future applications of superfine particles.

Attention is now being paid across a broad spectrum of industry to the possibility of using superfine particles as a way of extending the existing range of new materials. The two research projects outlined above have clearly had a substantial influence in turning superfine particles from university research specimens into objects of potential interest to industry.

4.3.2 Gas Evaporation Method

A piece of transparent glass held around 10 cm above a lighted taper will first start to cloud on its underside and the taper will eventually become invisible when viewed through the glass. If the taper is then extinguished and the underside of the glass inspected it will be found to be covered by a black, sooty film. If this sooty film is scraped off and rubbed between the finger and thumb it will be found to infiltrate even the smallest cracks in the skin, thereby furnishing clear evidence of the extremely small size of its component parts. In order to get a more immediate impression of its smallness it can easily be compared with the kind of flour found in any kitchen. If this is rubbed between the fingers in the same way then these particles (flour particles have a size distribution of between 10 and 100 μm) of flour will be found to be rather rough in texture when compared with the particles of soot. The size of the soot particles could be anything from a few micrometers down to submicrometer levels.

Taper soot can readily provide anyone with first hand experience of superfine particles. The flame of the burning taper also shares many common features with the gas evaporation method to be outlined below.

The generation of superfine particles by the gas phase method can easily be verified using the type of vacuum evaporator commonly found in the laboratories of any university engineering department.

During the course of the vacuum deposition process employed by this piece of apparatus the metal atoms which have been melted and evaporated from a tungsten heating boat travel directly through the high vacuum atmosphere, only rarely colliding with the residual gas molecules in the evaporating chamber, and eventually adhere to the underside of the substrate at the top of the chamber. The substrate is at this time heated to a temperature of around 300°C. As the vaporized atoms reach the surface of the substrate they form a thin film but as the base (substrate) has been heated this causes the accumulating layers of atoms to crystallize into a crystalline evaporation film. Evaporation films are currently finding applications in a broad range of industries and in particular in the electronics, electrical and optical fields (Fig. 4.2 illustrates the principle of vacuum deposition although it is not intended to suggest that such a mechanism could be applied immediately to an industrial scale process without some degree of modification).

The vacuum deposition mechanism outlined above can thus be used for the production of superfine particles. Figure 4.3 shows a number of requisite auxiliary parts such as a gas cylinder and regulator valves.

The method itself involves charging the tungsten heating boat with the evaporation material and then pumping out the air from the evaporation chamber (just as with vacuum deposition) until the vacuum has been increased to a high level in the order of 5×10^{-3} Pa[3] (3.8×10^{-5} Torr). The vacuum pump regulator valve is then closed and a mixture of inert gases such as Ar and He are pumped in until the pressure reaches the required level (for example, 1.3 kPa, 10 Torr) for superfine particle evaporation. The tungsten evaporation boat is then heated and when it reaches a temperature in excess of the melting point of the evaporation materials (it is not possible to generalize but this temperature could well be in the region of 200–300°C) smoke will begin to rise from the vicinity of the materials in the heating boat in much the same way as was observed in the area around the taper flame referred to above. The

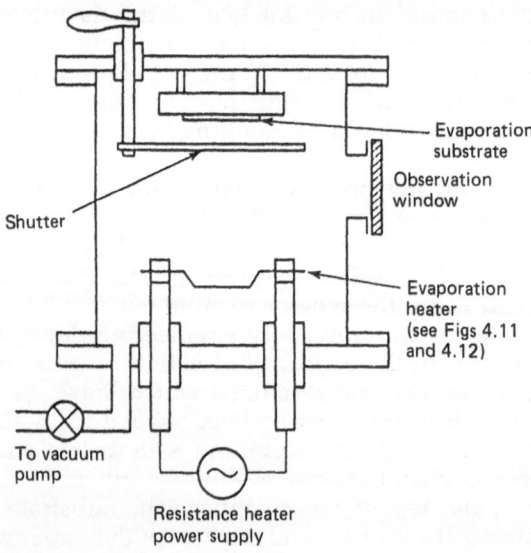

Fig. 4.2 Typical experimental vacuum evaporator.

[3] Pascal, 1 Pa = 7.6×10^{-3} Torr.

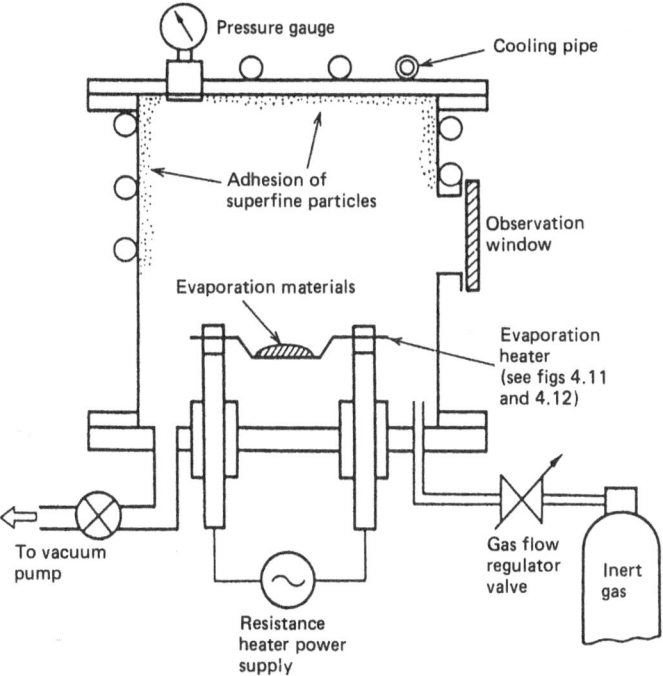

Fig. 4.3 An experimental device for the generation of superfine particles by gas evaporation using an experimental vacuum evaporator.

presence of superfine particles in this smoke is now a well-documented fact.

Shigeki Yatsuya (during research carried out in the Engineering Department of Nagoya University – now with the 3M Company in the USA) used a video cassette recorder to observe the conditions under which smoke containing evaporated metal atoms was given off. In order to do this he evaporated Mg in an atmosphere of He gas and found that the smoke form underwent numerous transformations as these conditions were changed [13].

The superfine particles generated inside a sealed evaporation chamber mix with the gas in the vicinity of the evaporation surface and rise towards the inside walls of the chamber to which they adhere in view of the comparatively low temperature (somewhere in the region of room temperature). On completion of the batch operation evaporation process the chamber is opened up and the outside air allowed to enter. The deposits on the inside walls can now be rubbed off with a brush and collected. The average deposit following a single evaporation operation (evaporation time of about 4–5 minutes) is in the order of several tens of milligrams with superfine particles sizes of less than 0.1 μm (100 nm).

The group to which the writer belongs has also carried out investigations into the process by which superfine particles are generated during gas evaporation [14]. Figure 4.4 is a schematic representation of the experimental method used. In this case, however, stationary state evaporation was continued for longer periods (several hours, for example) than was the case with the experimental vacuum deposition device described above. Under fixed evaporation conditions the superfine particles which have been generated can be recovered at any time during the course of the continuous evaporation process.

The metallic atoms which have evaporated from the surface of the molten metal are cooled by the surrounding gas molecules (which are much cooler than the vaporized atoms) thereby generating superfine particles by causing flocculation of the condensing atoms. Superfine particle size (the number of agglomerated atoms) can be regulated by controlling the conditions outlined above and small superfine particles with diameters in the 2 nm range[4] have been produced by this method.

This type of superfine particle generation has been investigated experimentally using the method illustrated in Fig. 4.5 [14]. In this particular piece of experimental apparatus, specimen collection rods have been set up at five different points between the evaporation area and the recovery area to enable the collection of particles at each point. In

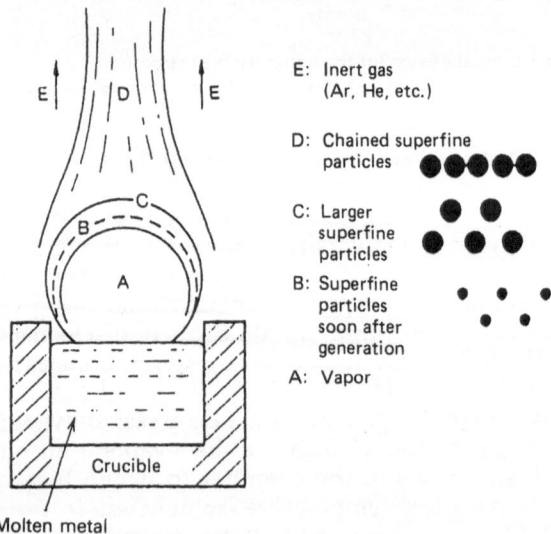

Fig. 4.4 Generation of superfine particles using the gas evaporation method (heating is by high frequency thermal induction).

[4] The boundary between superfine particles and clusters (referred to earlier) lies somewhere in this range.

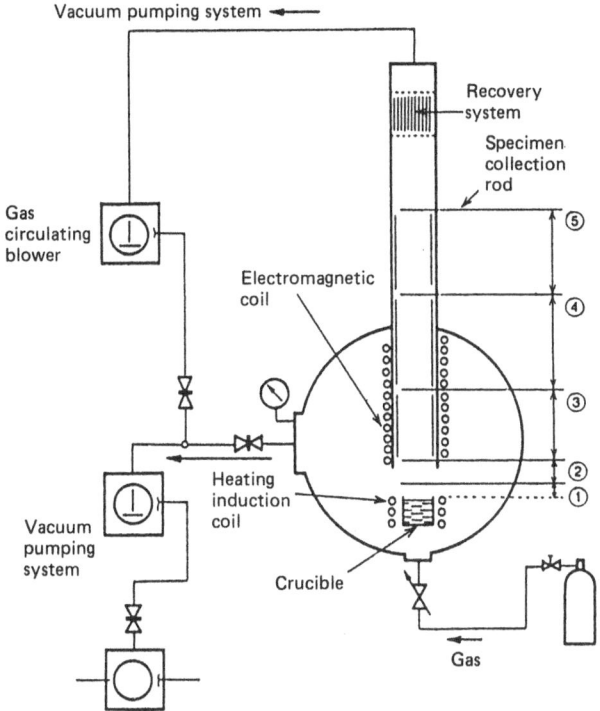

Fig. 4.5 Schematic representation of apparatus for the generation of superfine particles by the gas evaporation method (high frequency induction heating) [14]. Specimen collection rods 1–5 are used to recover superfine particles immediately after generation.

order to collect the particles in mid-flight the ends of the specimen collection rods are fitted with the type of mesh carbon film used in electron microscope photography. The ends of the rods are positioned in the vapor stream containing the superfine particles, the shutter opened momentarily and the rod then withdrawn again from the system to permit observations to be made with an electron microscope. Figure 4.6 shows a photograph of some superfine particles (composition: Fe–Co) recovered by collection rod 1, which is located in the lowest position, immediately after generation. By and large these particles range in size between 2 and 5 nm. Figure 4.7, showing particles collected by collection rod 2, clearly demonstrates that the particles have agglomerated by this point to form something between 10 and 100 superfine particles with diameters in the 20–30 nm range. No further growth (agglomeration) has been observed from this point onwards. Solenoids positioned both above and below specimen collection rod 3, as shown in Fig. 4.5, form a magnetic field with a strength of approximately 200 Gauss at its center. As shown in Fig. 4.8, the magnetic superfine particles (composition:

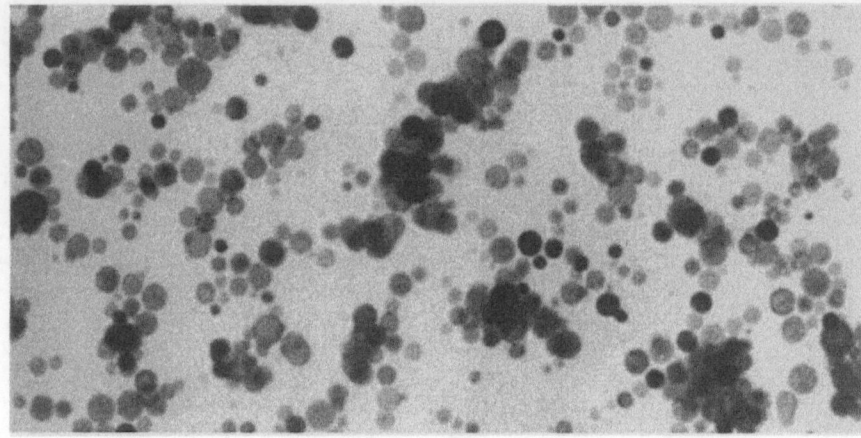

Fig. 4.6 Superfine particles immediately after generation by the gas evaporation method (particle size: 2–5 nm).

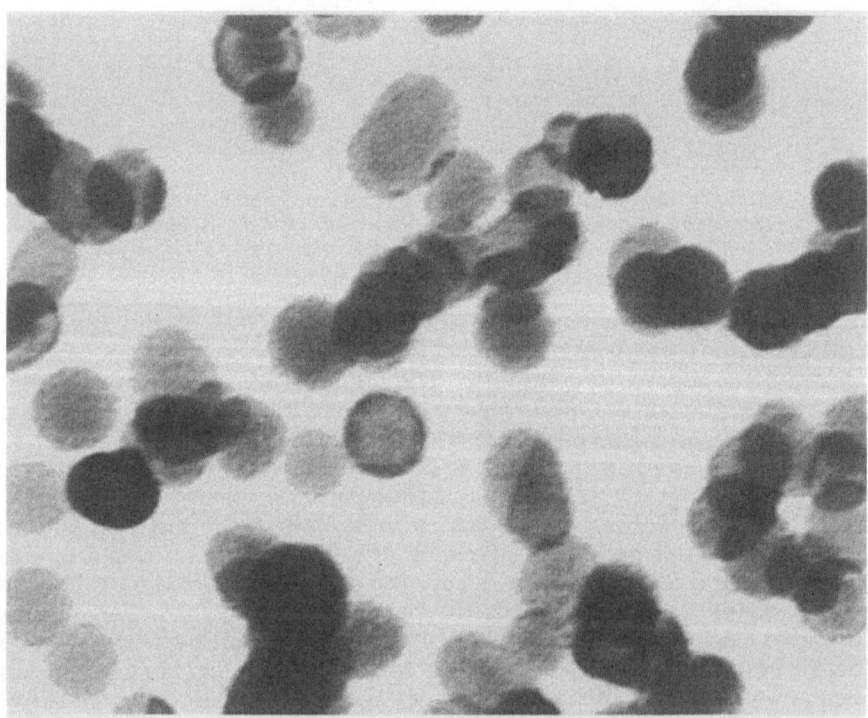

Fig. 4.7 Superfine particles formed by agglomeration of smaller particles soon after generation (particle size: 20–30 nm).

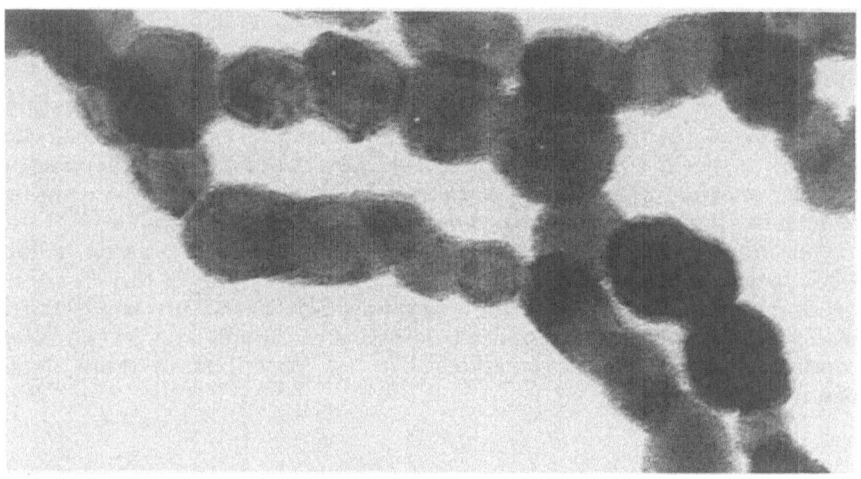

Fig. 4.8 Chain formation of superfine particles of magnetic alloys (particle composition: Fe–Co alloy, average size: 20 nm).

Fe–Co) passing through this magnetic field form into chains of particles. The particle size distribution shown in Fig. 4.9 is sharp and extremely uniform. It is, however, impossible to alter the sizes of superfine particles after they have been made although uniformity of size has become a

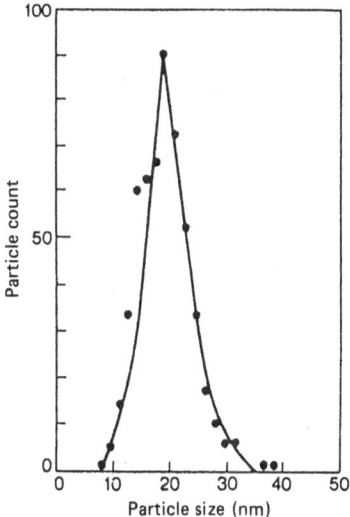

Fig. 4.9 Distribution of superfine particles of the metal alloy Fe–Co generated by the gas evaporation method (high frequency induction heating)[14].

required condition for certain types of application if the best possible use is to be made of the special characteristics of the type of particle in question.

The method of producing superfine particles outlined above (see Fig. 4.5) involves the use of high frequency heat for evaporation. In the production method which made use of an experimental vacuum deposition device, on the other hand, evaporation was achieved by means of resistance heat. A number of typical heating methods using gas evaporation, including the two referred to above, are illustrated in Fig. 4.10. The superfine particles which constitute the targets of these various heating methods, along with their individual characteristics, are detailed in Table 4.1. A number of these heating methods and evaporation conditions have been investigated and are described in more detail below.

a. Resistance Heating

The most commonly used evaporation heat sources are the coil filament or boat-shaped resistance heaters used in the vacuum deposition process. These methods are both illustrated in Fig. 4.11. Since evaporation materials such as W, Mo and Ta are normally suspended when in the coiled form or melted in a boat this sort of heating and evaporation method is not available where, for example, (a) the combination involves two materials (heating body and evaporation material) which will melt to form an alloy at high temperatures, or (b) the evaporation temperature of the evaporation material is higher than the softening temperature of the heating body. The main materials used these days for the evaporation process are metals with relatively low melting points such as Al, Ag, Cu and Au.

The resistance heater shown in Fig. 4.12 is a basket-shaped W filament coated with a refractory material such as Al_2O_3. The melted evaporation material does not therefore come into direct contact with the high temperature heating body material which means that it is possible to use a heated alumina crucible to evaporate materials such as Fe or Ni which have relatively high melting points (in the region of 1500°C) by comparison with the group of metals listed above.

The output of the heater is quite sufficient at about 1 kW although this will result in the production of just a few tens of milligrams of superfine particles through evaporation and deposition on the interior walls of the evaporation chamber of between 1 and 2 g of evaporation material. If the amount of superfine particles generated is to be increased then this can only be done by repeating the process, and the operation itself is therefore, as explained above, clearly no more than a research tool and does not have any direct industrial application as it stands. However, with the addition of just one small component (compare Fig. 4.2 and Fig. 4.3) the volume of superfine particles produced can be raised to a much more significant level.

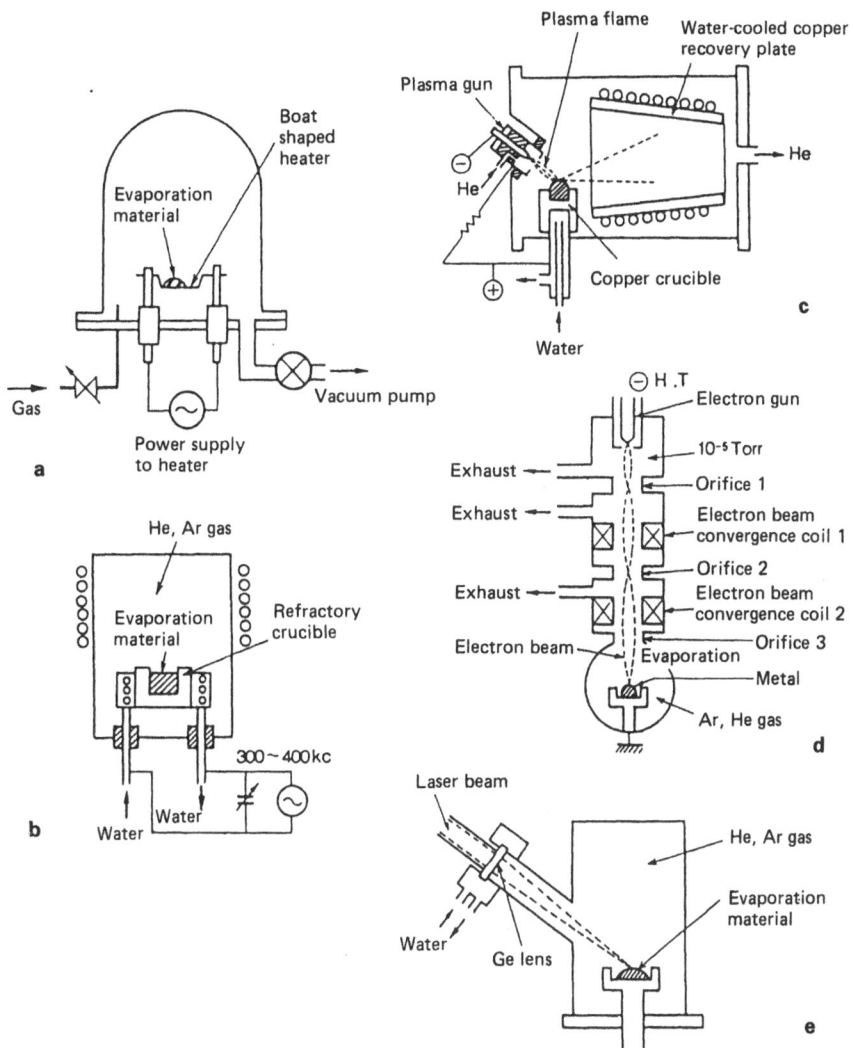

Fig. 4.10 Heating methods used in the generation of superfine particles by gas evaporation. a Resistance heater; b high frequency induction heater; c plasma jet heater; d electron beam heater; and e laser heater.

Table 4.1 Generation of superfine particles by gas evaporation

Name	Heating/evaporation method	Generation atmosphere	Special features
Resistance heater	Resistance heaters take a number of different forms such as the boat, filament or basket shape. Evaporation material is placed on the heater and heated to evaporate it	(inert) (reducing) 1.3×10^2–1×10^5 Pa	Easily created on a laboratory scale but the production volume per operation is only in the milligram order
Plasma jet	A plasma jet is used to heat the metal materials which are mixed together in a water cooled copper hearth	(inert) 2.6×10^4–1×10^5 Pa	Well suited to laboratory scale production (20–30 g/batch). Can be used with most metals.
High frequency induction	Metal materials are placed in a refractory crucible and heated by means of high frequency induction. There is a churning effect inside the crucible	(inert) 1.3×10^2–6.5×10^3 Pa	Particle diameter is easily controlled to ensure a fair degree of uniformity in terms of size. Long operation and high output is possible
Electron beam	Differential pressure is maintained by a slit between the high vacuum electron beam generation chamber and the evaporation chamber with a pressure of approximately 1 Torr. The raw material is fed in in the form of a wire	(inert) (reactive) 1.3×10^2 Pa	Enables the production of superfine particles of metals such as Ta or W or compounds such as TiN or AlN all of which have high melting points
Laser beam	The material is irradiated by a continuous, high energy light source (CO_2 laser, for example) which is concentrated into a beam and directed into the receptacle through a Ge window or lens	(inert) 1.3×10^3–1.3×10^4 Pa	The structure of the evaporation vessel is quite simple and can be used for the evaporation not only of metals but also of chemical compounds and minerals. It is also effective with compounds such as SiC

Table 4.1 Continued

Name	Heating/evaporation method	Generation atmosphere	Special features
Plasma arc sputtering	A d.c. voltage is applied across the gap between the evaporation material which forms the cathode and the ring shaped anode in an inert gas atmosphere. The electrical discharge melts the surface of the cathode thereby causing the material to evaporate	Mixture of both (inert) + (reactive) gases 1.3×10^4 Pa	Particle diameter can be controlled thereby permitting the maintenance of a degree of uniformity in terms of particle size
Direct electric current heating	An electrode in the form of a carbon rod is pressed down onto the surface of the evaporation material in its bulk solid form and an electric current applied. The heat generated by the passage of the current melts the surface of the evaporation material which evaporates as it burns the rod back	(inert) $5 \times 10^2 - 5 \times 10^3$ Pa	In addition to SiC this method can be used to produce superfine particles of other carbides such as Cr, Ti, V, Nb, Ta and W

(inert): inert gas; (reducing): reducing gas; (reactive): reactive gas

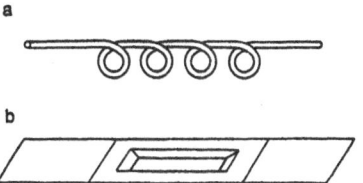

Fig. 4.11 Types of resistance heater used for the evaporation process. **a** Coil, and **b** boat.

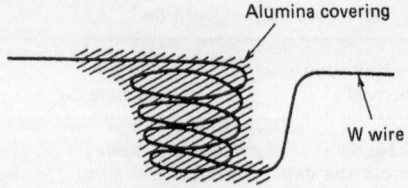

Fig. 4.12 Basket-shaped tungsten wire filament with alumina covering.

b. Plasma Jet Heater

Small quantities of superfine particles may be sufficient for the purpose of observing their form through an electron microscope or for determining their physical properties but at least 10 g is required if we are to make any meaningful assessment of their properties when in powder form. With this in mind Wada first tried during the late 1960s to make use of the plasma jet technique for the purpose of superfine particle evaporation. The technique had previously only been used in the melting and welding of metals [12,15]. The method is illustrated in Fig. 4.10(c) and the apparatus itself is shown in Fig. 4.13. For this operation first the metallic evaporation material is placed in a water-cooled copper crucible. A high

Fig. 4.13 Experimental apparatus for the generation of superfine particles by means of plasma jet heating.

frequency d.c. voltage is then applied across the gap between the evaporation material and a plasma gun mounted diagonally above the crucible. The effect of this is to ionize the inert gas (He or Ar, for example) flowing through the gun thereby causing the plasma to arc.

The atmospheric evaporation pressure inside the generation chamber is controlled by means of adjustments made to the flow of gas into the chamber and the exhaust gas discharged from the chamber via the vacuum system. If the force of the flow from the plasma gun is increased then so too is the volume of superfine particles generated, but if the intensity of the plasma flame is concentrated on just a part of the melt surface it is possible to observe, through the observation window fitted in the side of the generation chamber, a denser plume of smoke (the flow of gas containing the superfine particles) rising from the evaporation material, thus indicating a greater degree of evaporation. The superfine particles so generated adhere to a water-cooled copper cylinder and the carrier gas is discharged as exhaust from the chamber. After 10 minutes the operation is suspended, slow oxidation carried out, the generation chamber opened up and the superfine particles recovered from the inside surface of the copper cylinder. The bulky nature of the powder thus produced will ensure that there is now enough to fill several 1 liter receptacles. Figure 4.13 shows a typical piece of apparatus used for the generation of superfine particles by means of the plasma jet heating technique. The gas cylinder (capacity: 7 m^3) positioned on the left side of the apparatus will probably give the reader a clearer idea of its actual size.

The output of the plasma gun is in the region of 10 kW. This method is used, as illustrated in Table 4.2, to produce superfine particles of various metals including Ta which has a particularly high melting point (2996°C) [16]. The superfine particles in Table 4.2 are listed in order

Table 4.2 Generation of superfine particles of metals using the plasma jet heating technique

Type	Generation conditions				Generation rate (g/min)	Average particle diameter (nm)
	Pressure (Torr)	Voltage (V)	Current (A)	Input (kW)		
Ta	760	40	200	8	0.05	15
Ti	760	40	200	8	0.18	20
Ni	760	60	200	12	0.8	20
Co	760	50	200	10	0.65	20
Fe	760	50	200	10	0.8	30
Al	400	35	150	5.3	0.12	10
Cu	500	30	170	5.1	0.05	30

Operation time: 1.0–1.5 h
Gas: He + 15% H$_2$
Internal diameter of crucible: 30 mm (water-cooled copper crucible)

according to the melting points of the various metals. The rate of generation (in this case the ratio of superfine particles recovered to evaporation time) is low in the metals Al and Cu which have low melting points. This is because the evaporation material is heated by plasma jet in a water-cooled copper crucible and when melted (or half melted) it comes into contact with the water-cooled Cu walls. The thermal conductivities of Cu and Al are among the highest of all the metals which means that the heat loss to the crucible walls is higher, thereby dissipating a high proportion of the plasma gun input to the water-cooled copper crucible. Where, however, the melting point of the evaporation material is high there is also a heat loss due to thermal radiation from the evaporation surface in addition to the heat loss to the crucible walls. The heat loss from radiation is in direct proportion to the fourth power of the absolute surface temperature of the material and as a result its effect begins to be felt as the temperature rises past the 2000 K (roughly the melting point of Ti) mark. This effect is most clearly evidenced by the low generation rate of Ta. Nevertheless, the effect of the reaction of a metal with the crucible is generally speaking ignored and superfine particles of most metals are made in this way. One cycle (about 60 minutes) will normally yield several tens of grams of particles.

Table 4.2 clearly indicates that this method is best suited for the production of superfine particles of Ni and Fe (which belong to a group of transition metals whose alloys are used as magnetic materials). Figure 4.14 illustrates the effect of changing the plasma current during the course of superfine Ni particle generation. Generally speaking, there is a tendency for any increase in plasma jet current to be accompanied by a similar increase in the diameter of the beam. This effect manifests itself in the diameter of the superfine particles which are generated. In other words if a 100 A current is concentrated in a plasma beam then the current density will be increased and the overheating of the plasma spot will result in the generation of much larger particles. If on the other hand a smaller current is used then, although the plasma beam will continue to concentrate the current, the plasma output itself will be smaller, resulting in a lower melt temperature and smaller particles.

The use of this method (the water-cooled copper crucible method may be a more apt name than the plasma heating method) creates a temperature gradient across the melt surface (the hottest part being in the center of the plasma spot and the coolest in the peripheral areas closest to the walls of the water-cooled copper crucible) with the result that the diameter of the superfine particles generated is far from uniform. In addition, the plasma generating cathode (normally a thin W rod) and the cooled Cu nozzle, which is fitted to the tip of the plasma gun to focus the beam, must be designed in such a way that there will be no deformation resulting from prolonged operating times.

Solutions will have to be found to these problems before we can seriously consider the industrial application of this method of superfine

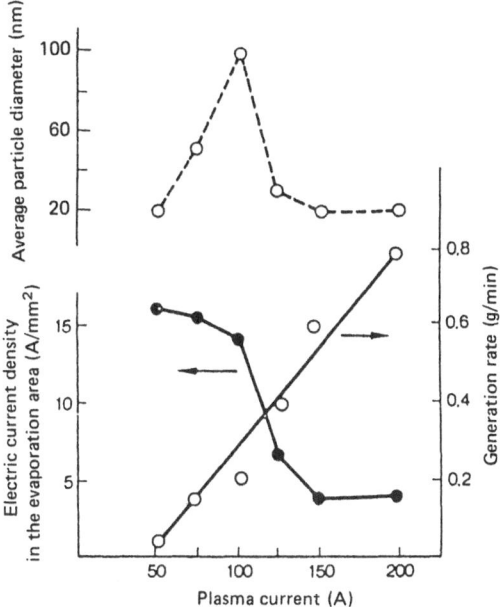

Fig. 4.14 Average particle diameter and generation rate of superfine particles of Ni using the plasma jet heating technique [16].

particle generation but the technology itself is in its infancy and there is still plenty of time for it to prove its worth.

c. High Frequency Induction Heater

In the early 1970s the Research Development Corporation of Japan entrusted Shinku Yakin (KK) with the development of the high frequency induction heater, a method which was identified during the development of techniques for the manufacture of superfine particles for use in high performance magnetic tape [17]. The use of the high frequency induction heating technique had had some notable successes in, for example, the vacuum melting of metals and was particularly attractive in the present circumstances. because (a) the melting temperature for evaporation can be kept constant, (b) the uniformity of alloys in the melt is good, (c) a steady operating output is possible over long periods, and (d) a megawatt heat source large enough for industrial production purposes had been used successfully[5] in vacuum melting operations. Points (a) and (b) result

[5] The largest high frequency induction heater currently in use for vacuum melting operations has an output of 2000 kW.

from the churning effect of the inductive action which causes the melt in the crucible to stir and thereby effectively prevents a temperature difference from developing between the central and outer parts of the evaporation surface. The continuing uniformity of the alloy in the crucible is also preserved in the same way.

Figure 4.15 is a schematic representation of the experimental device used to prove Hayashi's theory by means of which the concept of the high frequency induction heater was demonstrated [18]. Figure 4.16 reflects the data obtained during the generation of superfine Cu particles by heating and evaporating approximately 50 g of Cu in a small crucible 25 mm high with an internal diameter of 20 mm [16]. The diameter of the generated particles was controlled by adjusting the pressure of the evaporation atmosphere and the melt temperature (output of the heat source), a relatively simple operation by comparison with the plasma jet technique illustrated in Fig. 4.14. Gas type, say Ar or He, is another easily controllable element in this process.

The special feature of this heating method is that the larger the scale of the operation (the larger the crucible), the greater the tendency toward uniformity of particle size. When high frequency induction heating techniques are used to melt metals in a refractory crucible there is a concomitant magnetic side effect which generates a flow of melt both up and down and towards the periphery. This continuous stirring of the melt surface helps to keep the temperature uniform across the whole of the surface. The superfine particles pictured in the photographs in Figs.

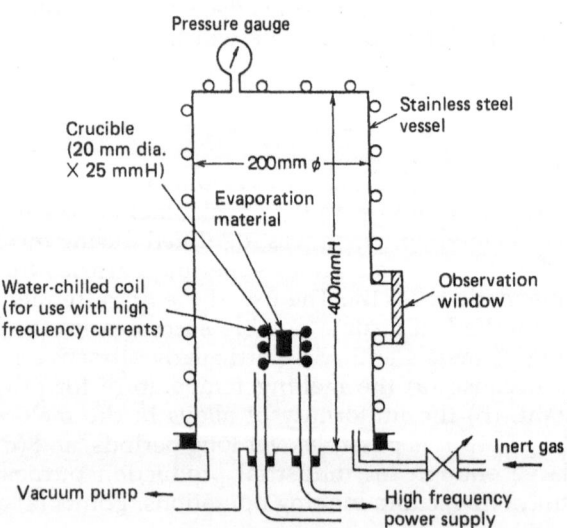

Fig. 4.15 Experimental device for the generation of superfine particles using a high frequency induction heating technique [16].

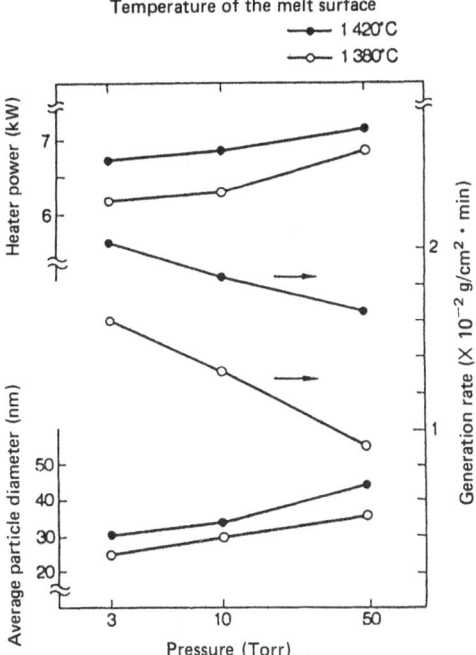

Temperature of the melt surface
—●— 1 420°C
—○— 1 380°C

Fig. 4.16 Conditions for the generation of superfine particles in an experimental device using high frequency induction heating [16].

4.6–4.8 were generated from an alloy of Fe–Co (magnetic) using a large high frequency induction heating device (Fig. 4.5) built specifically for the purpose of this research program. Taken in conjunction with the particle distribution shown in Fig. 4.9 these figures provide a clear illustration of the sort of particle size uniformity obtainable through the use of this method.

d. Electron Beam Heater

Electron beam heating is used in a variety of different operations such as melting, welding, deposition and precision machining and is normally carried out in a high vacuum. The emission of electrons from a cathode inside an electron gun involves the heating of the cathode surface in order to generate a stream of electrons which is in turn accelerated through a high vacuum (0.01 Pa or less) through the application of a high voltage to the beam. Even in subsequent electron beam systems the raising of the pressure above this level has resulted in the generation of abnormal electrical discharges within the beam generation system and in the repeated collision of the beam with residual gas molecules, thereby scattering the beam and preventing it from making effective contact with its target. For this reason, when electron beam heating is used in, for example, a melting operation, an extremely powerful vacuum pump is required in order to preserve the high level of vacuum inside the melting chamber.

This compares with a normal gas pressure level of around 1 kPa which is required in the evaporation chamber for the gas evaporation method. The development of electron beam heating techniques depended very much on the discovery of a solution to these conflicting pressure requirements and this was provided by Saburo Iwama (professor in the engineering department of Daido University).

Iwama introduced a number of orifices for the purpose of creating pressure differences at various points along the path of the electron beam from the gun where the high acceleration voltage is applied through to the evaporation chamber itself. A vacuum pump is used to intensify the vacuum at each of these gaps and at the same time the beam scatter is reduced through the use of an electronic lens to focus the beam and train it towards the evaporation chamber [19]. It has now been clearly established that the high calorific projection density of the electron beam as a heat source is particularly well suited for use in the evaporation of metals with high melting points such as W, Ta and Pt. However, the constant discharge of exhaust through the orifices creates a tendency for the generated particles to be drawn in the direction of the electron gun.

In order to combat this tendency Iwama introduced a gas inlet as shown in Fig. 4.17 immediately above the evaporation chamber in the

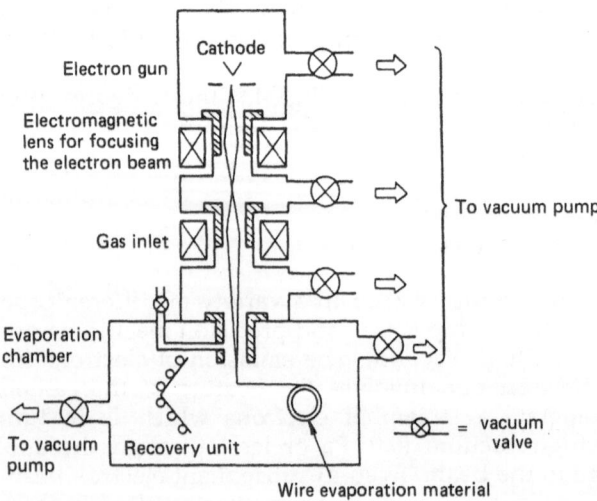

Fig. 4.17 Device for the generation of superfine particles by gas evaporation using an election beam heating technique [20].

differential area of the final orifice. Most of the gas injected via this inlet flows into the evaporation chamber thereby maintaining the required pressure for the generation of superfine particles. The design whereby the gas flows from the direction of the orifice area towards the evaporation chamber (towards the bottom of the diagram in Fig. 4.17) also produces a number of other beneficial effects in that (a) it prevents the generated particles from escaping into the electron beam system, (b) it eliminates contamination of the electron gun and electron beam system, and (c) it makes long periods of continuous operation possible.

Another innovation was the use of evaporation material in the form of a wire, the tip of which was introduced into the path of the electron beam as it entered the evaporation chamber and advanced continuously in accordance with the rate at which it was melted and evaporated [20]. In the evaporation of metals with high melting points such as W, Mo, Ta or Nb or of activated metals such as Zr or Ti (used in plasma jet heating) it is impossible to prevent a reaction between the molten metals and the crucible except by the use of a water-cooled crucible (there is currently no sign of the development of a crucible which does not react with the molten metals it contains). This experiment involving as it did the melting and evaporation of the evaporation material without the use of a crucible thus prevents the admixture of impurities originating in the interaction between metal and crucible.

Using the wire feed technique outlined above, an investigation of evaporation volumes was carried out using W wires of various diameters between 0.2 and 0.7 mm. The results of these tests, shown in Fig. 4.18, clearly demonstrate the increase in the volume of evaporation which accompanies a reduction in the diameter of the wire. The volume

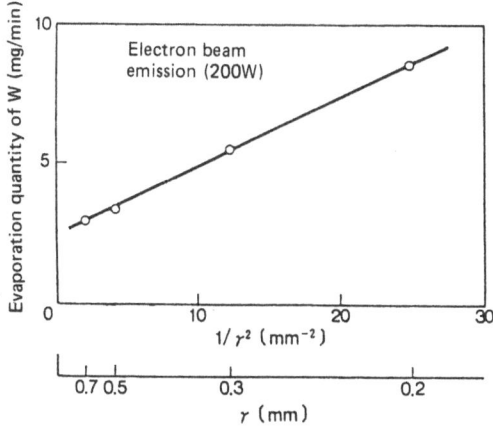

Fig. 4.18 Relationship between diameter and evaporation quantity of a W wire in the path of an electron beam [19].

decreases, in fact, in inverse proportion to the square of the wire diameter or, in other words, the wire section. Iwama also points out that the element of heat loss through the wire by conduction from the melting and evaporation point is considerable.

The gas injected into the evaporation chamber during the course of the superfine particle generation process is drawn off by the vacuum pump (an oil rotary type vacuum pump is used in view of the pressure level of approximately 100 Pa in the evaporation chamber). Superfine particles produced by the evaporation process are thus also drawn off as part of this flow of exhaust gas. The superfine particle recovery unit is positioned between the evaporation chamber and the discharge outlet and has a recovery rate in the region of 80%. This also demonstrates the relative ease with which the recovery rate can be increased by utilizing the flow of gas at the low 100 Pa pressure level.

Iwama has used the electron beam heating technique illustrated in Fig. 4.17 to generate superfine particles from various compounds with high melting points such as TiN and AlN. In each case a reactive gas was injected, N_2 during the evaporation of Ti to produce superfine TiN particles and NH_3 during the evaporation of Al to produce superfine particles of AlN. The superfine particles of TiN were found to have a size of 10 nm or less and a solid crystal habit, and the superfine particles of AlN measured in the region of 8 nm.

The utilization of an electron beam heating technique as part of the gas evaporation process is highly effective in the generation of superfine particles of metals and compounds with high melting points.

e. Laser Beam Heater

An optical heating technique which has found wide application in recent years is the laser and there are a number of special benefits which accompany its use as a heater in the generation of superfine particles. These are: (a) the heat source can be located outside the evaporation chamber thereby eliminating the effect of the chamber entirely, (b) any kind of material such as metals, compounds or ores can be melted and evaporated, and (c) the heat source (laser beam) remains uncontaminated by the evaporation material.

Figure 4.10(e) shows the basic design of a device for the generation of superfine particles using a laser beam heater. In much the same way as the resistance heater (Fig. 4.10(a)), this device is able to make use of the laboratory vacuum evaporator with the laser beam generated outside the system entering the chamber via a window containing a single crystal Ge or NaCl plate. The evaporation material can be held in position inside the chamber by means of a small refractory retainer (attention must still, of course, be paid to the potential for interaction between the retainer and the evaporation material at high temperatures). One experiment along the above lines was carried out by Wada when he irradiated ordinary SiC powder (a-SiC) with a CO_2 laser beam in an Ar gas atmosphere. The use of powder as the target material is, in fact, effective at the experimental stage when the laser beam output is only small. Beams with a greater output can be used to produce continuous evaporation by irradiation of bulk solid materials. As the atmospheric pressure is increased so too is the size of the particles generated. Superfine SiC particles generated in an atmosphere of Ar gas at 1.3 kPa have a diameter of approximately 20 nm [21]. The Si content of the SiC superfine particles was evaluated by means of X-ray diffraction intensity and was found to increase with the increase in gas pressure as shown in Fig. 4.19.

The problem, however, is the amount of light which can effectively be absorbed by the surface of the target materials, particularly metals, during the course of irradiation. Matsunawa (assistant professor in the welding and engineering research center of Osaka University) carried out experimental metal evaporation operations using an Nd:YAG laser (this laser uses neodymium ions Nd^{3+} and $Y_3Al_5O_{12}$ (Yttrium, Aluminum, Garnet) as its basic material), which has better metal surface photoabsorption properties than a CO_2 laser. This is a near infrared laser one an order of magnitude smaller than the CO_2 laser at 1.06 µm. The laser used in this piece of research was an Nd:YAG pulse laser with an average maximum output in the region of 200 W, a pulse width of 3.6 ms and an irradiation energy capacity of between 20 and 33 J per pulse. Pulse irradiation by this laser in an atmosphere of inert gas such as He has been used to generate superfine particles of metals such as Fe, Ni, Cr, Ti, Zr, Mo, Ta, W, Al, Cu and Si [22].

Matsunawa has used this same irradiation technique in a reactive gas phase to generate superfine ceramic particles such as oxides and nitrides. The size of the superfine particles generated can be regulated by varying the pressure of the atmosphere during evaporation just as with the other gas evaporation techniques.

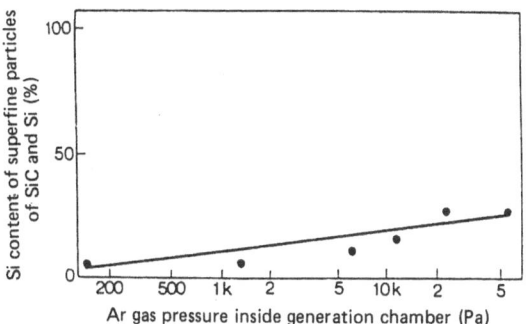

Fig. 4.19 Relationship between gas pressure and the ratio of superfine Si particles to Si/SiC particles generated using a laser beam heat source [21].

Akinobu Yoshizawa (former assistant professor in the engineering department of Tokyo University) focused the output of a TEA-CO_2 laser in a gas phase, thereby generating a strong electric field in the vicinity of the focal point. The gas phase chemical reaction induced by the generated pulse plasma was used in the generation of superfine particles [23]. The output of the laser was in the region of 0.5 J per pulse and the superfine SnO_2 particles generated were single crystals of just a few nanometers in diameter. The particle distribution itself was comparatively sharp.

Lasers with large outputs are, of course, already being used for industrial purposes and increased efficiency of laser energy absorption by material surfaces would be a most welcome development in view of the close relationship of this technology with the evaporation and generation of superfine particles.

f. Recent Innovations in Gas Evaporation Techniques

The gas evaporation technique has thus far been targeted primarily at the production of superfine particles of metals but superfine particles can, in fact, also be generated from inorganic and organic chemical compounds (such as ceramics or polymers) and from composite metals.

1. *Superfine Oxides.* Figure 4.20 shows a device containing two solid Al (purity: 5 N) electrodes in an atmosphere of inert gas such as Ar containing traces of O_2. The generation of an arc discharge between the two electrodes causes the surface of each to melt and superfine particles to be generated. According to Tsukasa Hirayama (at the time a participant in the Hayashi Superfine Particle Project and now with Nippon Denso (KK)) conditions where heating takes place by electric arc discharge in an atmosphere consisting of a mixture of 40 kPa of Ar gas and 13 kPa of O_2 gas are sufficient for the generation of superfine particles of Al_2O_3 [24].

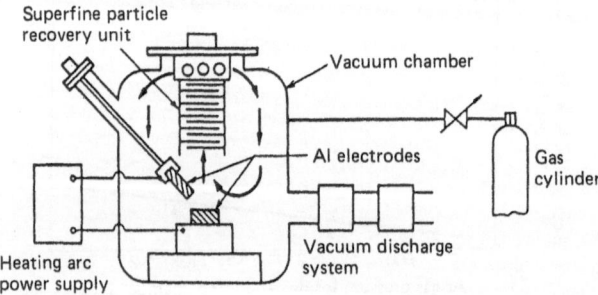

Fig. 4.20 Device for the generation of superfine particles of γ-Al_2O_3 by means of electrical arc discharge.

The superfine particles of Al_2O_3 had an outstanding crystalline structure and heat resistance characteristics such that the γ-Al_2O_3 particle shape scarcely altered during continuous heating at 1260°C over a period of 1 hour. Superfine particles of γ-Al_2O_3 are well suited to use at high temperatures as media for precious metal catalysts with clusters of Pt or Rh precipitated on their surfaces.

2. Superfine Particles of Organic Compounds. The comminution of organic and polymer compounds has traditionally been achieved either by milling from a material in its bulk solid form or by polymerization in a liquid phase as typified by the production of fine latex particles. Hideki Toyotama (formerly with the Hayashi Superfine Particle Project and now with Stanley Denki (KK)), however, decided to experiment with the use of a standard gas evaporation technique in order to achieve the same goal. In a number of different experiments he succeeded in melting and evaporating bulk organic compounds in Ar gas to produce superfine particles varying in size from a few tens of nanometers up to several micrometers [1,25]. Figure 4.21 shows a photograph of superfine particles of pyrene taken with the help of a fluorescence microscope. The size of the particles can be controlled by regulating the pressure of the Ar gas atmosphere during the evaporation process. The particle distribution is also reasonably sharp as illustrated in Fig. 4.22.

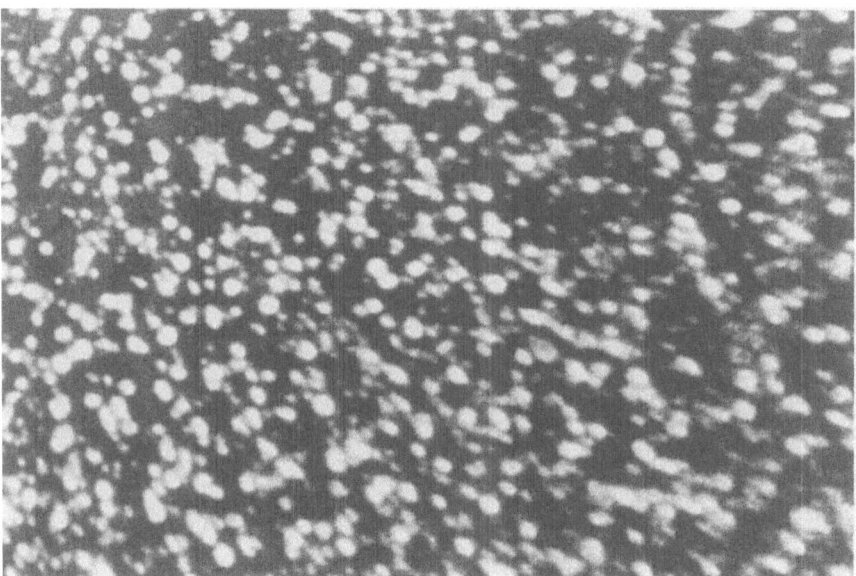

Fig. 4.21 Fluorescence microscope photograph of superfine particles of the organic compound pyrene [25]. Average particle size: 300 nm. (Picture by Hideki Toyotama.)

An interesting feature of superfine particles of organic compounds is that compounds which exhibit water repellent, hydrophobic characteristics in their bulk solid form have been found to disperse readily in water in their particulate form thus exhibiting a corresponding increase in surface hydrophilic characteristics. Of late the pharmaceutical companies have been showing considerable interest in this phenomenon in view of its potential application in the field of hydrophobic vaccines. This represents a completely new departure in the application of superfine particle technology.

3. Superfine Particles of Composite Metals. It was held for a long time that, in the generation by gas evaporation of compound superfine metal particles from elements exhibiting fundamentally different vapor pressures, the fact that the same evaporation source was used made the composition of the final product difficult to control (double digit vapor pressure differences have been controlled at the evaporation temperature under experimental conditions). Masaaki Oda (formerly with the Hayashi Superfine Particle Project and now with Shinku Yakin (KK)), however, designed an original technique for feeding the evaporation material through and thereby succeeded in the generation of composite superfine particles from a combination of Cu and Zn, two metals with very different vapor pressure characteristics [26].

This method involves the feeding of a Zn rod into the crucible containing the Cu melt just as it starts to evaporate. Composition control is achieved by regulating the temperature of the molten Cu and the rate at which the Zn rod is fed in.

Superfine particles generated in this way have a composite form consisting of a superfine particle Cu core of some several tens of

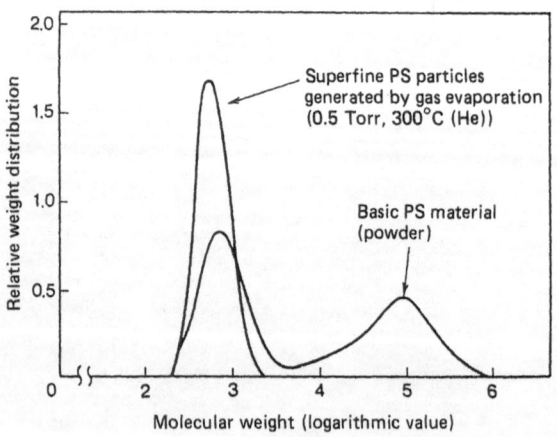

Fig. 4.22 Molecular weight distribution of superfine particles of polystyrene (PS) generated by gas evaporation [25].

nanometers in diameter covered by a layer of fine ZnO crystals some 2–3 nm deep. These particles have outstanding activation and selectivity characteristics when used as catalysts in the synthesizing of methanol.

4.3.3 Active Hydrogen–Molten Metal Reaction Method

Masahiro Uda (formerly of the Science and Technology Agency, National Research Institute for Metals and now with Nisshin Seiko (KK)) formulated the theory which formed the basis for his research into the generation of superfine metal particles by this method, while analyzing the diffusion and emission of the gases N_2 and H_2 in molten metals during the course of arc welding operations using a plasma mixture containing hydrogen gas as the heat source.

The notable feature of this method is the fact that superfine particle generation increases as the concentration of hydrogen gas in the plasma is increased. As an example of this phenomenon Uda succeeded in generating superfine particles of pure iron at a rate of 20 mg/s using a concentration of 50% H_2 gas in Ar gas, an arc voltage of 30–40 V and an arc current of 150–170 A [27]. Uda also demonstrated the possibility of producing a high degree of uniformity in the size of the particles generated by causing them to travel as quickly as possible away from the arc heat source. The superfine particles were recovered by separating them from the carrier gas using a combination of a centrifuge recovery unit such as a cyclone and a filtration recovery unit such as a filter. Superfine ceramic particles were generated in the form of TiN and AlN by evaporating Ti and Al using a nitrogen plasma heating technique, the result being confirmed by means of an X-ray diffraction test [28]. Uda is currently entertaining high hopes for the application of superfine ceramic particles generated using the chemical effects produced in these types of oxidizing, nitriding and deoxidizing atmospheres.

4.3.4 Sputtering Method

Film formation by sputtering[6] in a vacuum is gaining rapidly in popularity as a result of the recent growth in demand from the semiconductor industry for such compound film, and is now one of the most powerful physical film production (PVD) techniques in use. The technique itself differs from other PVD techniques such as vacuum deposition or ion plating in that the evaporation material (known as the target) is not heated to melting point during the course of film formation in order to generate the phenomenon known as evaporation whereby the surface atoms of the evaporation material become detached.

---

[6] Sputtering: ions generated by glow discharge in an Ar gas atmosphere bombard the surface of the target cathode, causing atoms to fly off.

This feature of sputtering attracted Yatsuya's attention and led him to attempt the generation of superfine particles by this method. The principle advantages of the method are: (a) there is no need for a crucible to hold the melt, (b) the evaporation material (target) can be placed in any position (either upright or facing downwards), (c) superfine particles can also be generated from materials with high melting points, (d) the evaporation surface area can be extensive, (e) compound particles can be generated by reactive sputtering (using reactive gas), and (f) the formation of thin films of superfine particles is possible.

Yatsuya experimented with this technique in a number of different forms, one of which is illustrated in Fig. 4.23. In this experiment Yatsuya placed two metal plates (an Al plate forming the anode and the evaporation material target plate forming the cathode) opposite each other in parallel in an atmosphere of Ar gas (40–250 Pa). An (acceleration) voltage of several hundred volts d.c. was then applied across the gap between the plates to generate a glow discharge. The ions in the glow discharge between the two electrodes bombard the cathodic evaporation material target causing atoms to evaporate from the surface of the target. The controlled generation of superfine particles is achieved by varying the discharge voltage, current and pressure. In the case of Ag as the target, the particles generated were between 5 and 20 nm in size and the evaporation rate obtained was found to have a virtually linear relationship to the size of the target's surface area [29].

Hiro Oya (Suzuka National College of Technology) on the other hand has succeeded in generating superfine particles by sputtering in a high pressure environment as distinct from the method employed by Yatsuya (the Yatsuya method is known as the arc plasma method) [30]. This method involves raising the target to a high temperature to melt the surface. The method is illustrated in Fig. 4.24 and consists of cathodic

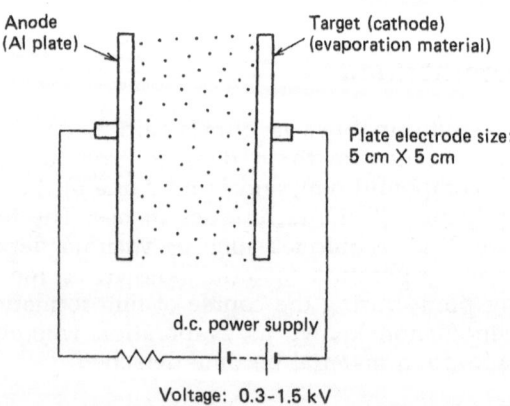

Fig. 4.23 Principle of superfine particle generation by means of the sputtering technique [29].

evaporation material in the shape of a ring which is set along with the corresponding anode in a mixed atmosphere of $15 H_2$ + He gas at a pressure of 13 kPa. A d.c. voltage is applied to generate an electrical discharge which melts the surface of the evaporation material (target) thereby generating the desired evaporation effect. The superfine particles generated immediately after evaporation travel through the hollow anode in the upper part of the chamber in order to reach the surface to which they adhere.

The superfine particles generated can be sufficiently well controlled to ensure an average particle size in the 10–40 nm range. This method is one of the preferred methods where particle size uniformity is an important consideration. In the case of an average particle size of 11 nm, for example, the size distribution curve indicates that roughly 90% of all the particles fall within 50% of the peak value. At the same time a generation rate of 50 mg/min-kW was achieved in the case of Fe, 34 mg/min-kW in the case of Cr and 28 mg/min-kW in the case of Ag.

This method is attracting some attention at this current stage in the research and development of particle generation technology in view of the potential it offers for the generation of superfine particles of composite materials by combining different elements (metals and chemical compounds) to form the evaporation target materials.

4.3.5 Method of Vacuum Evaporation to Running Oil Surface (VEROS)

In the earliest stages of thin film formation by deposition in a high vacuum it has been observed that bodies of a fairly uniform size roughly

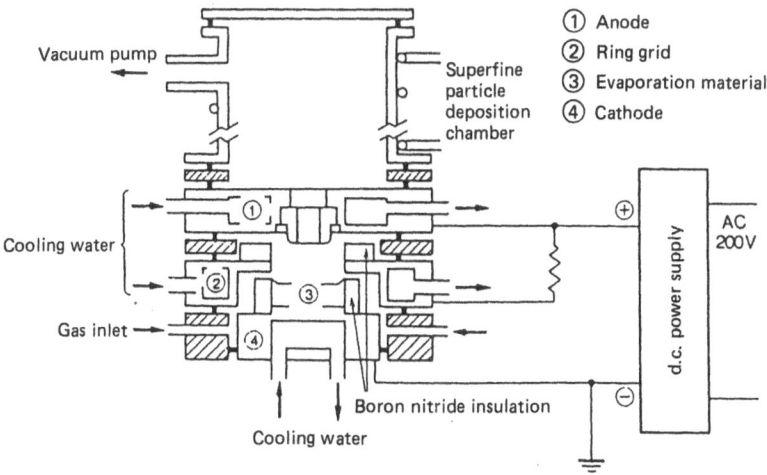

Fig. 4.24 Generation of superfine particles by the arc plasma sputtering method [30].

equal to that of superfine particles tend to adhere to the surface of the substrate. This phenomenon is known as island formation [31]. If evaporation is continued beyond this point, however, these isolated bodies tend to join up to form a thin film at first and later a thick film.

Yatsuya was prompted to make use of this phenomenon by evaporating a metal on to a substrate well smeared with oil and then recovering the oil and along with it the superfine particles created by the agglutination of the metal atoms in the oil. The experimental device which he used is illustrated in Fig. 4.25 [32].

The evaporation material is placed in a water-cooled copper crucible where it is heated and evaporated in a high vacuum by means of an electron beam heater. When the required conditions for evaporation are satisfied an overhead shutter is opened to permit deposition of the evaporated material on the underside of a rotating disk. At the same time oil is pumped down through a hole in the center of the disk whereupon it is dispersed towards the perimeter by centrifugal force. This results in the formation of a thin fluid oil film along the underside of the disk which is spun off into receptacles positioned around the walls of the chamber.

The types of oils used are those with low vapor pressures such as silicon oil which is used as a pumping fluid in the oil diffusion pump. The disc itself rotates at between 200 and 400 rpm. The superfine particles residing in the oil film are recovered along with the oil in receptacles housed around the perimeter walls of the evaporating chamber. The

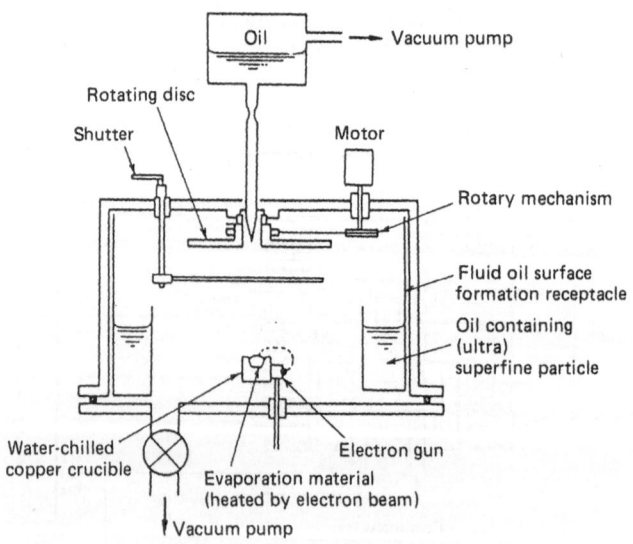

Fig. 4.25 Generation of (ultra) superfine particles by vacuum deposition on to a fluid oil surface (VEROS) [32].

proportion of particles contained in the oil, however, is only small and the mixture has therefore to be concentrated by vacuum distillation to reduce it to a paste. This method has been used in the generation of superfine particles of, for example, Ag, Au, Pd, Cu, Fe, Ni, Co, Al and In.

Special features of the VEROS method include: (a) the average particle diameter is in the region of 3 nm (it is not so easy to generate such small particles by the gas evaporation method), (b) the particles generated exhibit a fair degree of uniformity in terms of size (see Fig. 4.26), and (c) the superfine particles are isolated and well dispersed in the oil from the outset.

Particle size can be controlled by varying the evaporation conditions. The conditions for generating large sized particles include: (a) a high evaporation rate, (b) a high level of oil viscosity, and (c) a small number of disk rotations. It is also possible to exercise some control over particle size by regulating the temperature of the superfine particles in the oil (by keeping them within the 100°C–150°C range). The use of any of the above control methods will result in an average particle size of 3–8 nm [33].

The use of the VEROS method results in the production of a paste containing superfine particles and is one of the methods used for the creation of superfine particles in an isolated state (with diameters of 5 nm or less).

4.3.6 Method of Evaporation Using Direct Electrical Current Heating

Yoshinori Ando (assistant professor in the Department of Science and Engineering of Meijo University) first started experimenting with this

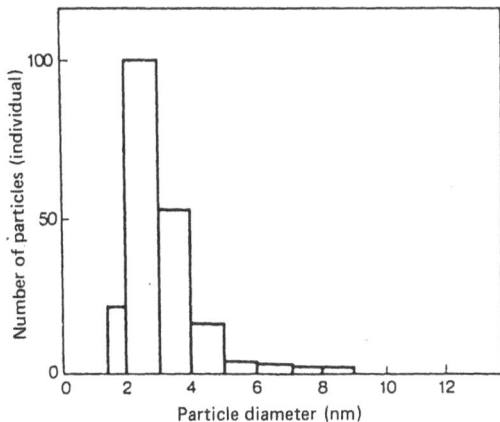

Fig. 4.26 Distribution by size of superfine particles of Ag generated by the VEROS method.

method in an attempt to generate superfine particles of SiC, one of the new ceramic materials on which a lot of hopes are pinned [34]. Figure 4.27 shows a typical piece of apparatus using this method, the central feature of which is a carbon rod electrode through which a current is passed as it is simultaneously pressed down against the upper surface of the Si (evaporation material) in bulk solid form. The atmosphere consists of the gases Ar or He (used in gas evaporation) at a pressure of between 1 and 10 kPa. Si is, of course, highly resistant to the flow of electric current at low temperatures and will not conduct current in this state. If, however, it is heated in advance from underneath then the resistance to the flow of current through the Si plate will be greatly reduced. The device thus makes use of the semiconductor characteristics of Si in that when it has been heated sufficiently to produce a substantial reduction in the level of electrical resistance an a.c. current of several hundred amperes is then passed through the material. The following phenomena can now be observed: (a) as the current continues to flow the carbon electrode first becomes red hot and then white hot, (b) as the carbon rod is pressed down onto the Si, the Si melts and eats away the surface of the rod, and (c) smoke is emitted with considerable force from

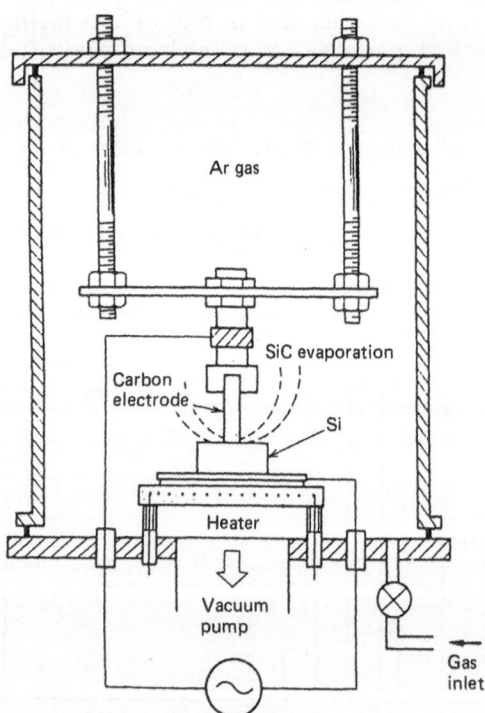

Fig. 4.27 Generation of superfine particles of SiC by gas evaporation using direct electrical current as the heat source.

around the carbon rod (as the temperature rises to levels in excess of 2200°C). If the current flowing through the carbon rod is increased then so too is the amount of smoke generated.

A current of 400 A in an atmosphere of Ar gas at 400 Pa has enabled recovery of superfine SiC particles at a rate of around 0.5 g/min. The ratio of Si to SiC is governed by the evaporation temperature as it relates to the vapor pressure of each of the two elements. The differences between the atmospheres created by the different inert gases Ar and He are also reflected in the nature of the superfine SiC particles generated. Generally speaking an Ar gas atmosphere is conducive to the generation of large particles with crystal habit whereas an He gas atmosphere is more conducive to the generation of smaller, spherical particles.

In addition to the generation of superfine SiC particles as described above, Ando used this same technique of heating the evaporation material by means of an electric current passed through a carbon electrode to generate superfine crystalline particles from various carbides such as Cr, Ti, V and Zr. He also succeeded in generating superfine, amorphous particles from metals such as Hf, Mo, Nb, Ta and W (all metals with high melting points). The origin of this phenomenon is thought to lie in the fact that the metals in the latter group have higher melting points than that of the carbon rod (electrode) and are therefore not fully melted during the above process. The smoke which is given off thus contains amorphous, superfine particles of carbon [35].

4.3.7 Hybrid Plasma Method

Toyonobu Yoshida (professor in the engineering department of Tokyo University) has devised a (hybrid) heating method which combines d.c. plasma (an independent heating method used in the generation of superfine particles as shown in Fig. 4.10(c)) with the kind of RF (inductively coupled) plasma[7] which is already used in industry. This method, which is illustrated in Fig. 4.28, involves the generation of RF plasma by coupling a high frequency magnetic field of several MHz from an induction coil wound on a quartz tube. The advantages of the RF plasma method include: (a) the plasma is non-polarized and thus resists the intrusion of polarized impurities (during melting and evaporation, for example), (b) reactive gas can be used, and (c) the plasma space is large and the flow rate of the gas through the space is slow by comparison with the d.c. plasma method which means that substances can be delayed longer in the plasma space (there is ample scope for the heating and reaction of a substance).

One disadvantage, however, with the use of RF plasma in the generation of superfine particles is the fact that the RF plasma flame is easily

---

[7] A 1 megawatt torch is currently being used.

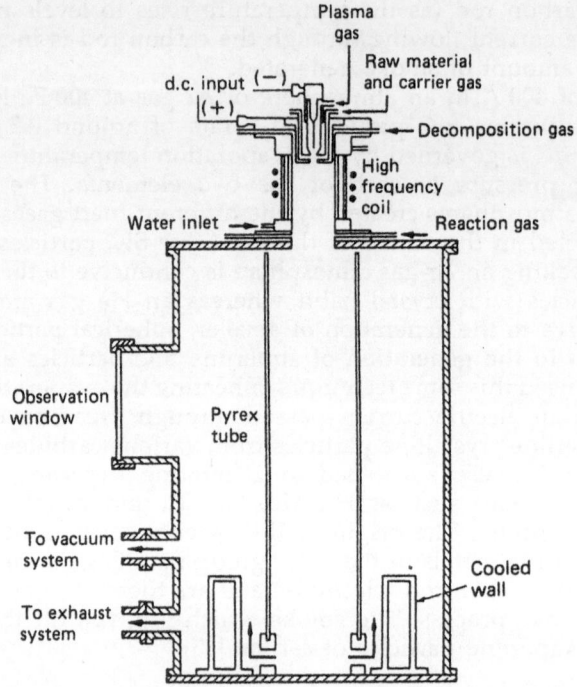

Fig. 4.28 Device for the generation of superfine particles using a hybrid plasma heat source [36–38].

disturbed by the intrusion of any gas or other material into the plasma space (particularly if entry is from a radial rather than an axial direction). Yoshida solved this problem by introducing a d.c. plasma arc jet which was aimed axially through the plasma space to act as a kind of pilot flame to sustain the RF plasma flame. The apparatus shown in Fig. 4.28 is designed around this concept [5,36–38].

The following are just three of the methods used for the generation of superfine particles by using plasma space for heating, evaporation and reaction:

1. The plasma evaporation method: generates superfine metallic particles by creating a combined stream of metal and gas.
2. Reactive plasma evaporation method: generates superfine particles of compound materials by adding a reactive gas to a stream of metal and gas such as that described in 1. above.
3. Plasma CVD method: generates superfine particles of compound materials by adding a reactive gas to a stream of compound material (gaseous) and gas and then discharging the excess gas from the system.

Of the above, the plasma CVD method, for example, has been used in the generation of superfine particles of Si_3N_4. The procedure adopted here was to inject the basic material Si_3N_4 (in fluid form) at a rate of approximately 4 g/min into a hybrid plasma space where it was thermally decomposed by the separate injection of H_2 gas. The quenching reaction produced by the NH_3 gas in the flame tip resulted in the generation of superfine particles of Si_3N_4. The diameter of the particles was 30 nm or less, the amorphous white nitrogen content was between 30 and 37 wt%, the Si content was in the region of 60 ± 2 wt%, and crystallization took place at an extremely high 1550°C which clearly demonstrates the high degree of purity obtained.

Yoshida rates the hybrid plasma method very highly in the generation of a whole variety of different superfine particles and is now looking to the refinement of the various hardware aspects in the shape of, for example, technology for stabilizing plasma and at the same time boosting its output in order to make the method suitable for industrial application in the future.

4.4 Outlook for Further Technological Developments in the Production of Superfine Particles

A wide variety of different techniques are currently being used for the production of superfine metallic and ceramic particles. However, most of these techniques are only in the research stage with many centers still concentrating on determining the characteristics of superfine particles and only just beginning to sense the possibility of identifying more practical applications for their work. The most likely pattern of development in the foreseeable future is: (a) the production of powders to facilitate the examination of the characteristics of superfine particles, (b) the production of powders for the development of practical applications for superfine particles, and (c) the development of techniques for the manufacture of superfine particles (or rather particulate powders) on an industrial scale. If this pattern is to be followed then the superfine particles (powders) produced will have to satisfy the following conditions:

1. They will have to meet the requirements of the users in terms of characteristics and quality.
2. Their quality will have to remain stable over prolonged periods of continuous operation.
3. The manufacturing equipment used will have to be capable of being scaled up geometrically in much the same way as current chemical plant.

For example, the conditions for the use of superfine particles or powder as a magnetic medium with high density magnetic recording characteristics are (a) a uniform particle size of several tens of nanometers, (b) a uniform metal alloy ratio which is capable of being adjusted, and (c) a chained form (where each particle constitutes a single crystal magnet). The generation of superfine particles by the gas evaporation method using the high frequency induction heating technique has already proved successful in meeting the various user requirements outlined above, such as qualitative stability during prolonged periods of continuous operation [39].

It is difficult at the present juncture to make any kind of definitive statement from the industrial point of view about the relative merits of the various methods of producing superfine particles (powders) outlined above. What we can say, however, is that there is currently a fully operational industrial installation (pilot plant) capable of producing tons of superfine particles by the month using a gas evaporation technique [40].

4.5 Handling of Superfine Particles

Superfine particles are extremely small particles with extremely large surface areas, close to the limit for a substance in its bulk solid form. For this reason superfine particles of a substance have characteristics which differ from those of the same substance in its bulk form. It is precisely these characteristics, encompassed within what are termed the volume effect and the surface effect, that researchers are attempting to exploit in their current work.

Following generation and before they are brought into contact with the outside air, superfine particles must first be slow-oxidized in order to prevent the rapid oxidization which would otherwise take place. Superfine particles have an extremely large surface-area-to-volume ratio and fresh surface particles must therefore be treated immediately after generation to prevent a rapid rise in the temperature of the surface layer due to heat generated by the oxidation process resulting from contact with air (oxygen).

Once treated, particles can be handled in the open air, but there remain a number of problems which cannot easily be dealt with using conventional powder (particle diameters in excess of several micrometers) handling techniques. For example, they are bulky, not free flowing, hard to compact (density does not increase even after formation), difficult to mix with other superfine particles or with larger size particles, and can only be adequately mixed with the help of large amounts of (liquid) binder.

It is therefore extremely important that suitable ways of handling superfine particles be well understood by those involved. However, a considerable amount of expertise is required and it would not be easy for us to do full justice to the vast number of different requirements for each individual industrial sector here. There are indeed a number of processing operations which are recognized to be indispensable when dealing with superfine particles. The most important of these are the range of surface treatments and processing techniques designed to enable the particles to be processed within controlled atmospheres (without bringing them into contact with the outside air).

The utilization of a technique which enables superfine particles to be processed into the required form without exposing them to the outside air can solve a large number of the problems which relate to their handling and at the same time permit the best use to be made of their special physical characteristics. With this objective in view, a variety of different methods have been used in the formation of superfine particle film [41–44]. The gas deposition method proposed by Hayashi is illustrated in Fig. 4.29. This method involves the introduction of the superfine particles into a stream of gas immediately after generation. This mixture is then ejected through a thin nozzle in the form of a jet (1–0.1 mm in diameter) and impacted against a substrate on which it forms a fine film (similar to vacuum deposition film) as shown in Fig. 4.30 [2,45,46].

The gas deposition method is an example of one of the newer techniques for the handling of superfine particles. The gas evaporation method which is used to create the required supply of superfine particles offers a number of advantages: (a) pure superfine particles can remain suspended in the gas phase within which they were generated, (b) the gas pressure at which the particles have been generated remains suitable

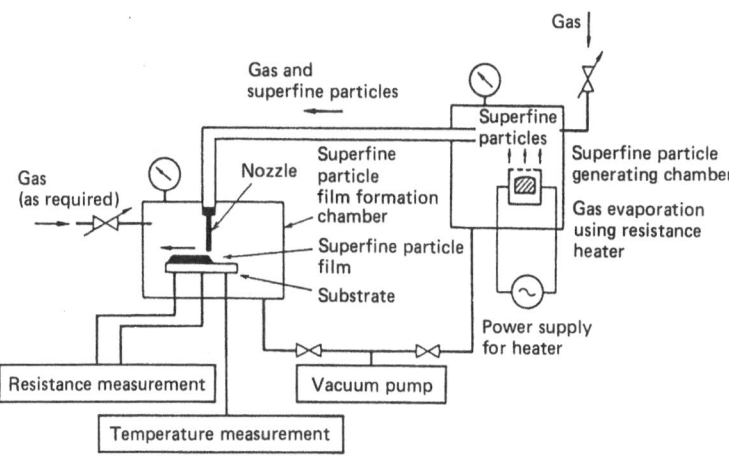

Fig. 4.29 Schematic representation of an experimental device for the formation of film by gas deposition [2,45,46].

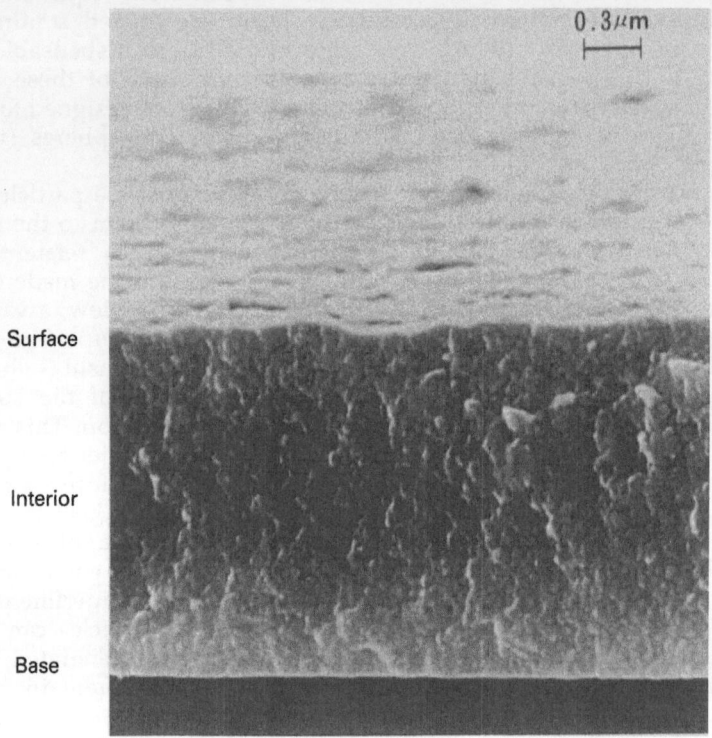

Fig. 4.30 SEM photograph showing a section through a film of superfine Cu particles formed by the gas deposition technique [2,46]. The superfine Cu particles were generated by means of gas evaporation.

for their further conveyance, and (c) the actual gas within which the superfine particles were generated can frequently be used unaltered to carry them further through the process.

The gas deposition method is an example of a technique which is related to the production of superfine particles and which is clearly developed with a view to the subsequent application of the particles produced. It is confidently expected that there will be many more instances of this type of development in the foreseeable future.

References

1. H. Toyotama: 35th Koubunshi Gakkai Nenji Taikai Kouen Yokoushuu, 35, 3, p. 763 (1986) (in Japanese)

2. C. Hayashi: Aerosol Kenkyuu, 1, 1, p. 23 (1986) (in Japanese)
3. The results of the Souzou Kagaku Gijutsu Suishin Jigyou Hayashi Cho-biryushi Project are given in the collection of papers (in Japanese) for the Hayashi Cho-biryushi Research Report at the Research Report Conference held by the Shin Gijutsu Kaihatsu Jigyoudan (December 1986). A general summary of the results is given in C. Hayashi: J. Vacuum Sci. Technol., A5(4), p. 1375 (July/Aug. 1987)
4. C. Hayashi: Ouyou Butsuri, 50, 2, p. 178 (1981) (in Japanese)
5. K. Akashi: Cho-biryushi Kagaku Sousetsu, 48, p. 29, Gakkai Shuppan Center (1985) (in Japanese)
6. S. Iijima: Tokushuu Cho-biryushi Nihon no Kagaku to Gijutsu, 25, 227, p. 29, Nihon Kagaku Gijutsu Shinkou Zaidan (1984) (in Japanese)
7. R. Uyeda: Ouyou Butsuri, 50, 2, p. 174 (1981) (in Japanese)
8. R. Kubo: "Cho-biryushi Kotai Butsuri Bessatsu Tokushugou", p. 5, Agne Gijutsu Center (1975) (in Japanese)
9. R. Uyeda: Nihon Kesshou Gakkaishi, 46, p. 19 (1974) (in Japanese)
10. K. Kimoto: "Cho-biryushi Kotai Butsuri Bessatsu Tokushugou", p. 46, Agne Gijutsu Center (1975) (in Japanese)
11. A. Tasaki: Jpn. J. Appl. Phys., 4, p. 707 (1965)
12. N. Wada: Jpn. J. Appl. Phys., 8, p. 551 (1969)
13. Yatsuya, Uyeda: Ouyou Butsuri, 42, p. 1067 (1973) (in Japanese)
14. E. Fuchita, M. Oda, S. Kashu: Proceedings of the 7th International Conference on Vacuum Metallurgy, p. 973 (1982)
15. N. Wada: Kinzoku, 48, 1, p. 50 (1978) (in Japanese)
16. S. Kashu, M. Nagase, C. Hayashi, R. Uyeda, N. Wada, A. Tasaki: Proceedings of the 6th International Vacuum Congress, p. 491 (1974)
17. R. Uyeda: Nihon Kinzoku Gakkaihou, 17, 15, p. 403 (1978) (in Japanese)
18. C. Hayashi: Japanese patent No. 793472 (1975)
19. Iwama, Asada: Seimitsu Kikai, 48, 2, p. 248 (1982) (in Japanese)
20. S. Iwama, K. Hayakawa, T. Arizumi: J. Crystal Growth, 56, 2, p. 265 (1982)
21. Uyeda, Kumasawa, Katou, Wada, Matsuda: Toyota Kenkyu Houkoku, 26, p. 66 (1973) (in Japanese)
22. Matsunawa, Katayama: Nihon Kinzoku Gakkai Fall Taikai Abstracts, p. 286 (1985) (in Japanese)
23. Sunouchi, Sakashita, Yoshizawa: Nihon Kinzoku Gakkai Fall Taikai Abstracts, p. 285 (1985) (in Japanese)
24. T. Hirayama: Shin Gijutsu Kaihatsu Jigyoudan Souzou Kagaku Gijutsu Suishin Jigyou Hayashi Cho-biryushi Project Research Report, p. 22 (1986) (in Japanese)
25. H. Toyotama: Kinou Zairyou, 7, 6, p. 44 (1987) (in Japanese)
26. Oda, Hayashi: 23rd Funtai nikansuru Touronkai Abstracts, p. 122 (1985) (in Japanese)
27. M. Uda: Nihon Kinzoku Gakkaihou, 22, 5, p. 412 (1983) (in Japanese)
28. M. Uda: Cho-biryushi no Jitsuyou Gijutsu, p. 39, CMC (1984) (in Japanese)
29. Uchiyama, Kamakura, Yatsuya, Mihama: Funtai Funmatsu Yakin Kyoukai 1985 Fall Taikai Abstracts, p. 168 (1985) (in Japanese)
30. H. Oya, T. Ichihashi, N. Wada: Jpn. J. Appl. Phys., 21, 3, p. 554 (1982)
31. Shinkuu Jouchakuhou de Hakumaku no Keisei to Kouzou nitsuite no Kaisetsu. For example: Kinbara, Fujiwara: Ouyou Butsurigaku Sensho, Hakumaku, p. 42, Shoukabou (1980) (in Japanese)
32. Yatsuya, Akao: Kotai Butsuri, 12, 4, p. 231 (1977) (in Japanese)
33. S. Yatsuya: Nihon Kinzoku Gakkai Kaihou, 18, 5, p. 377 (1979) (in Japanese)
34. Y. Andou: Ouyou Butsuri, 50, p. 153 (1981) (in Japanese)
35. Y. Andou: Shinkuu, 23, 7, p. 319 (1980) (in Japanese)
36. T. Yoshida: Tokushuu Cho-biryushi Nihon no Kagaku to Gijutsu, 25, 227, p. 35, Nihon Kagaku Gijutsu Shinkou Zaidan (1984) (in Japanese)
37. K. Akashi: Nihon Kinzoku Gakkai Symposium – Ceramic Cho-bifun no Seizou Process to sono Butsuri Kagaku paper, p. 1 (1981) (in Japanese)
38. T. Yoshida: Nihon Kinzoku Gakkai Symposium – Ceramic Cho-bifun no Processing paper, p. 1 (1985) (in Japanese)

39. Imanishi, Inoue, Ono, Miyatake: National Technical Report, 25, 1, p. 153 (1979)
40. M. Oda: "Cho-biryushi Kotai Butsuri Bessatsu Tokushugou", p. 103, Agne Gijutsu Center (1984) (in Japanese)
41. Abe, Hayakawa: Denshi Zairyou, 19, p. 79 (1980) (in Japanese)
42. Adachi, Okuyama, Kasaka, Tanaka: Aerosol Kenkyuu, 1, p. 123 (1986) (in Japanese)
43. S. Iwama, K. Hayakawa: Jpn. J. Appl. Phys., 20, p. 335 (1981)
44. H. Komiyama: Funtai to Kougyou, 19, 5, p. 22 (1987) (in Japanese)
45. S. Kashu, E. Fuchita, T. Manabe and C. Hayashi: Jpn. J. Appl. Phys., 23, p. L900 (1984)
46. S. Kashu: Funtai to Kougyou, 19, 5, p. 37 (1987) (in Japanese)

Additional References

47. Chikara Hayashi: "Ultrafine particles", Phys. Today, Dec., pp. 1–8 (1987)
48. K. Hatanaka, M. Kaito, M. Umehara, S. Kashu, C. Hayashi: "Preparation of superconducting thick films of Y–Ba–Cu–O by gas deposition of ultrafine powder", Proceedings of the 1st International Symposium on Superconductivity (ISS'88), pp. 341–345 (1988)
49. C. Hayashi, R. Uyeda, A. Tazaki, (eds.): "Cho-biryushi – Sozo, Kagaku Gijutsu", Mita Shuppankai (1988) (in Japanese)
50. M. Koishi (ed.): "Cho-biryushi Ouyou Kaihatsu Handbook", Science Forum, (1989) (in Japanese)
51. Chemical Engineering Tokushu: "Hakumaku, Cho-biryushi Zairyo eno Kitai", 33, 2, pp. 113–153 (1988) (in Japanese)

Further Reading

The following publications cover the topic of superfine particles in general:

"Kotai Butsuri Bessatsu Tokushugou Cho-biryushi" Agne Gijutsu Center (1984) (in Japanese)
"Nihon no Kagaku to Gijutsu – Tokushuu Cho-biryushi", Nihon Kagaku Gijutsu Shinkou Zaidan (1984)
"Kagaku Sousetsu No. 48 Cho-biryushi – Kagaku to Ouyou", Gakkai Shuppan Center (1985) (in Japanese)
Cho-biryushi Ouyou Gijutsu, Nikkan Kogyo Shinbunsha (1986) (in Japanese)

Chapter 5

Chemical Manufacturing Processes

The development of the microcomputer has had an enormous impact on both society as a whole and on industry in particular. Its influence has been felt in the field of materials technology, in that the range of material functions has been considerably increased and as a direct result the use of these materials has also inevitably increased. For this reason the reliability of materials has become a matter of some considerable concern. The principal characteristics of a material depend not so much on its physical as on its chemical properties. We have so far, however, concentrated almost exclusively on the physical properties of materials insofar as they have a direct connection with production methods and have made little mention of their chemical properties or of the relative advantages and disadvantages of the various different production methods from the point of view of chemical composition. It has, however, become abundantly clear that our objective of high material reliability depends in large measure on chemical properties such as the chemical composition of particles, the compositional changes which take place either within particles or at particle level and the secondary phase physical forms of particles. In this chapter we have focused our attention on those methods of producing fine particles which make use of chemical reactions at the various stages of the production process and also on the chemical composition and structure of particles generated by these various methods.

5.1 Introduction

A substance can normally take any of three basic physical forms: the solid form, the liquid form or the gas form. The atoms of which substances are composed hold fairly static positions in relation to each other when in their solid form, whereas in liquid form they are able to move around much more freely, and in gas form they are completely free to move without being influenced by neighboring atoms. This particular factor is explained by the free volume theory of substances. This theory holds that the free volume of a solid is low and its density correspondingly high with interaction between atoms remaining strong over long distances. A gas, on the other hand, has a high free volume with the interaction between atoms remaining weak, thus permitting each individual atom to move without influence from neighboring atoms. The properties of liquids place them by and large mid-way between solids and gases although in terms of the free volume theory they are much closer to the solid end of the range. The relative freedom of movement of atoms within a substance has a considerable bearing on the production of fine particles and on the characteristics of the particulate which is generated.

The production of fine particles is just one division of synthetic materials technology and chemical particle production methods consist in the main of those methods which depend for their effect on chemical reactions. Chemical reactions depend on changes in a substance's atomic structure and these changes are in turn tightly bound up with the substance's existential form. In other words, those features which are of particular importance when considering the minimum reactive unit of a solid are the size of the solid itself, the fact that the reactive area is restricted to the relatively small surface area of the solid and the fact that the phase of the substance generated has a strong influence on the continuation of the reaction. A further problem is the need for the application of large amounts of thermal energy in order to raise the temperature of the substance sufficiently to cause the atomic movement required for a reaction to take place. The reaction which takes place in the high temperature contact area necessarily produces a high level of agglutination between the two reacting substances and an increase in size of the system. The harder the material becomes, the more difficult it is to break it down into smaller units, of course. This brings with it a requirement for the introduction of various milling techniques for the comminution of substances in their solid form, which must be considered a negative factor when considering the most appropriate methods for the production of fine particles. Fine particle production processes using liquids and gases, on the other hand, benefit from the fact that the minimum reactive unit is to be found on the atomic or molecular level. At this level the reaction can move freely throughout the atomic or molecular system thus extending to the whole of the substance in question. Another advantage over solids is that fine particles synthesized in liquids form particulate solids during the generation phase which then precipitate out of the liquid, leaving the way clear for the continuation of the generation process. In liquids and gases therefore fine particles are produced by the build-up or accumulation of individual atoms and molecules. This in turn enables the effective control of particle size, shape and size distribution, which makes it altogether more appropriate as a way of producing fine particles. From the point of view of reaction space the biggest difference between liquids and gases is in their density. Liquids have a density of approximately 1 g/cm^3 whereas gases have a density in the order of 1×10^{-3} g/cm^3. In other words, a gas requires 10^3 times as much reaction space as a liquid to produce the same system of reaction. The production of fine particles in liquid rather than gas phase is particularly advantageous in that the higher density of liquids results in correspondingly higher rates of particle growth, lower facility costs, higher productivity levels and greater ease of maintenance. Figure 5.1 illustrates the particular features of solid, liquid and gas reactions.

The production of fine particles by chemical means implies that the production process itself engenders some kind of chemical reaction. Table 5.1 shows the range of chemical fine particle production techniques dealt with in this chapter, ordered in terms of the forms of the various systems of reaction and the methods by which the particulates are formed. In more general terms, chemical reactions may be taken to include such processes as the melting and solidification of pure substances. Processing methods which are based on such simple physical changes do not, however, form part of the subject matter of the present chapter. In other words, the production of metallic particulates by cooling vaporized metals has been included in Chap. 4, which dealt with physical methods of producing fine particles. However, processes such as atomization, where the basic particle production method is physical but where a significant part of the process depends on a chemical reaction, have been included in the present chapter. In such cases we have done our best to distinguish the chemical from the physical aspects of the processes involved. As pointed out above, chemical processes can take place both in a pure solid, liquid or gas phase and also in a mixed system of reaction. However, simply classifying the type of reaction which occurs between two solid phases as a mixed solid phase reaction was found to be unsuitable as a means of classifying fine particle production techniques. This is because the level on which solids can be blended is very much a macro level and there is also a requirement that some kind of milling technique be introduced in order to achieve the comminution of the particles formed by the reaction in order to produce the ultimate fine particulate. We have therefore based our selection of solid reaction fine particle production techniques on the extent to which the reaction through which the solid blend is created takes place on the level of the atoms which constitute the principal structural elements of the fine particles produced. For this reason particle formation by precipitation from a liquid phase may also be included. A liquid phase is, in fact, a very suitable medium for a chemical reaction and related chemical production techniques for fine particles are currently the principal focus of researchers' attention [1]. In the case of gaseous phase reactions the method for evolving the reaction gas in many cases is by a physical

Condition of reaction substance		Particular features
Solid	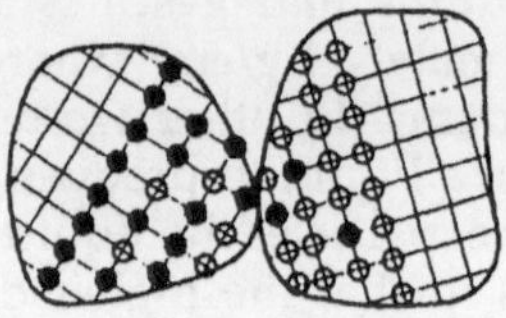	The atoms of a solid occupy relatively fixed positions and if there is to be a reaction involving changes in these positions then either the interatomic gaps must be crossed or else some mechanism must be found to enable the atoms to change places.
Liquid — Melt	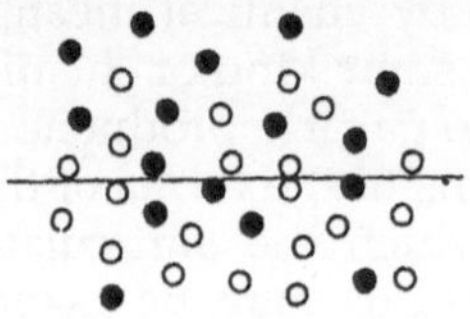	The density of the substance remains similar to that of the same substance in its solid form but the positions of the atoms are not fixed. Agitation of the melt can thus produce mixing at the molecular level.
Liquid — Solution	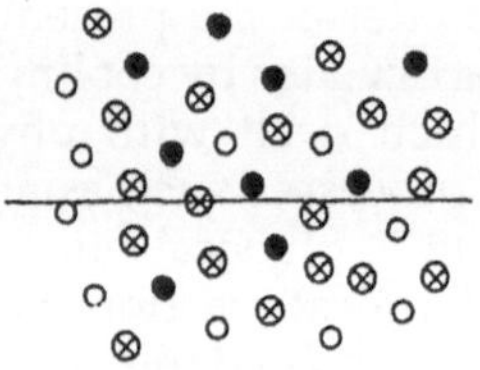	A solution consists of particles of a solid reactive substance suspended in a liquid phase. The dilution of the reactive substance with a solvent makes the reaction itself easy to control.
Gas	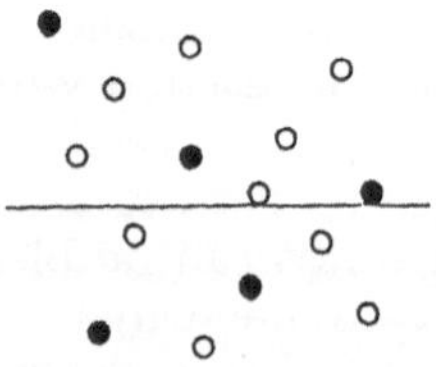	Atomic density is extremely low which means that there is a large amount of space which remains unused during a reaction. For this reason it is not easy to create nuclei around which fine particles can form. Normal practice is therefore to prepare solid reaction points in advance of the reaction itself.

Fig. 5.1 Typical expressions and features of solids, liquids and gases.

process such as vaporization. Here the gaseous phase reaction is for the manufacture of fine particles where the production of particles from the reaction gas depends upon a chemical process. We present as an example the use of a laser as the driving force of the reaction.

Table 5.1 Forms of reaction systems used in the production of fine particles by chemical means and mechanisms by which particulates are formed

Production method		Fine particle formation mechanism	System of reaction
Precipitation	Coprecipitation	Chemical	Liquid–liquid
	Compound precipitation	Chemical	Liquid–liquid
Hydrolysis	Inorganic salt hydrolysis	Physical	Liquid–liquid
	Alkoxide hydrolysis	Chemical	Liquid–liquid
Atomization	Atomized hydrolysis	Physical	Liquid–gas
	Atomization and calcination	Physical	Solid–solid
Oxidation–reduction	Liquid phase oxidation	Chemical	Liquid–liquid
	Hydrothermal oxidation	Chemical	Solid–liquid
Freeze drying		Physical	Solid–solid
Laser synthesis		Chemical	Gas–gas
Spark discharge		Physical	Solid–liquid

5.2 Precipitation

In the precipitation method a solution is prepared containing the metal atoms of the ceramics, and a suitable precipitator is added to this solution to bring about precipitation of the ceramic precursor, which is then subjected to calcination to obtain the fine ceramic particles. The precipitate is normally separated out from the solution which contains it by some kind of filtration technique and it is preferable, therefore, that the precipitate should not be difficult to filter. When the ion products $[A^+][B^+]$ of the A^+ and B^- ions in an aqueous solution reach a point where they exceed the solubility product, they will begin to bond and eventually form a crystal lattice. As the crystals gain in size they begin to form a sediment due to gravitational action and thereby precipitate out [2]. By and large, particles with diameters in excess of 1 μm will be subject to this kind of sedimentation. Particle growth leading to precipitation sometimes takes place around a single nucleus but more commonly takes place by virtue of secondary aggregation on top of a small primary particle. By increasing the size of the primary particle's diameter this kind of development facilitates the filtration process. The size of the particles which precipitate out of a solution depends on the relative rates of nuclear formation and nuclear growth. In other words, if the rate of nuclear formation is slower than the rate of nuclear growth

then the number of particles generated will be small whereas the size
the individual particles will be relatively large. However, the process
particle generation by means of precipitation is complex and no simp
way of controlling the relative rates of nuclear formation and growth h
yet been found. Generally speaking, experimental results suggest th
the lower the solubility of the precipitate, the smaller the size of tl
precipitated particles. Low levels of solution supersaturation, howeve
tend to result in comparatively large particle sizes. Control of tl
precipitation reaction is, in fact, rather difficult and so regulation
particle size is generally achieved by increasing the size of the precipita
itself. This method depends on the heating of the solution in order
age the precipitate which it contains. Since small particles are mo
soluble than large ones a solution becomes supersaturated more quick
in respect of larger particles. If, therefore, the temperature of a giv
solution is increased, thereby also increasing its solubility level, a
providing a reasonable time is allowed for the process to take place, th
it will be found that the smaller particles can be eliminated from t
solution and the larger particles increased in size.

The precipitation method has traditionally focused on the particle si
and form of the precipitate. Recently, however, the emphasis has shift
towards uniformity of composition. In other words, it has becor
necessary to find ways of controlling composition at particle level
order to reduce the size of the materials used, to carry out low temperatu
production and to exercise precise control of the microstructure
materials. The most commonly used methods of precipitation in t
production of fine particles are the coprecipitation method and t
compound precipitation method [3,4].

5.2.1 Coprecipitation

Coprecipitation is the name given by analytical chemists to a phenomen
whereby the fractional precipitation of a specified ion in a solution resu
in the precipitation not only of the target ion but also of other io
existing side by side in the solution. The additional precipitation
unwanted ions is, of course, an impediment to the analytical process.
the field of fine particle production, however, the method whereby
the batched ions in a solution are fully precipitated is known
coprecipitation.

It is, of course, possible to use the solubility product as the basis
a quantitative discussion of the occurrence of precipitation during t
course of coprecipitation in accordance with the theory of chemi
equilibrium. Some of the substances most commonly used in precipitati
operations are hydroxides, carbonates, sulfates and oxalates. pH
obviously an important element in hydroxides but even in cases such
oxalic acid, for example, where members of the hydroxyl group do r
enter directly into the precipitation process, pH is still an extreme

important element in view of the strong effect which the dissociation process has on it.

Conditions for precipitation to take place within a solution vary in accordance with the metallic ions present, and this forms the basis for the analytical chemist's use of the ion separation process. As far as fine particle synthesis is concerned, however, this is one of the major drawbacks of coprecipitation. In other words, it is fair to say that there are no metallic ions which precipitate under identical conditions and it is not therefore possible to induce simultaneous coprecipitation of the various ions which constitute a given material. As the pH rises the metallic ions in a solution precipitate in order one after another as the conditions for their precipitation are met, thereby creating either a pure or a mixed precipitate composed of one or more metallic ions. Let us examine, for example, the synthetic production of stabilized zirconia by coprecipitation with yttria, magnesia and lime. Figure 5.2 illustrates the relationship between the ion concentration and the pH in an aqueous solution of zirconium ions and stabilizer ions. Chlorides of the metallic elements zirconium, yttrium, magnesium and calcium are easily dissolved in water to form the basic aqueous solution. In order to produce fine particles of zirconia containing yttrium as the stabilizer, a base such as sodium hydroxide or aqueous ammonia is added to the solution containing the zirconium and yttrium ions, to induce the precipitation of ions from the solution. However, the simultaneous precipitation of both types of ions and the consequent generation of fine particles with a uniform composition at the particle level is very difficult to achieve. Figure 5.2 shows clearly the considerable difference between the precipitation of the zirconium and stabilizer ions. Fractional precipitation

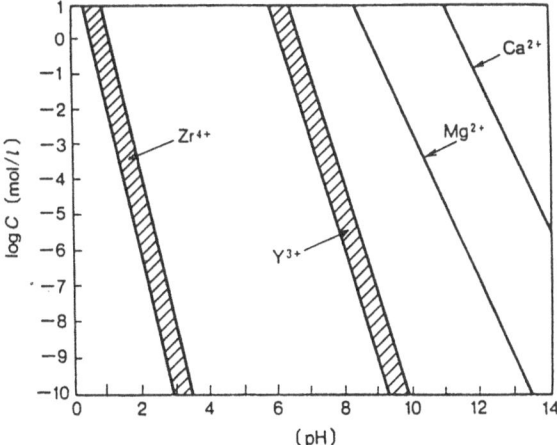

Fig. 5.2 Relationship between pH and concentrations of zirconium and various stabilizer ions in an aqueous solution.

finally results in the generation of a mixture of fine particles of hydrated zirconia and fine hydroxides of the stabilizer element. The resultant precipitated mixture can then be calcined to produce compound fine particles but, although the blending of zirconium and stabilizer atoms may have been more thorough than that which is achieved by a mechanical blending of the powdered oxides of the different components, it nevertheless remains the case that the basic reaction was solid phase, and the blend is difficult to stabilize at particle level.

In order to prevent this basic tendency to fractionate during precipitation it is possible to increase the concentrations of precipitators, such as sodium hydroxide or aqueous ammonia, introduced into a metallic salt solution, and thereby meet the precipitation conditions for all the metallic ions in the solution at the same time. To avoid localized changes resulting from precipitation in the newly introduced parts, the solution is churned vigorously during the course of the precipitation process. In this way it is possible to prevent a certain amount of fractionation during precipitation but there is no guarantee that uniformity of composition will be retained when the precipitate is heated to form the desired compound. For coprecipitation, the yardstick for the even distribution of microcomponents as part of the principal component is a difference of three or less in the precipitated pH or p (ion concentration) of the metallic ions which have participated in the precipitation process. In coprecipitation, the uniform distribution of components in the precipitate may be thought to have reached the particle level when the number of metallic atoms in each of the target fine particles is proportionally approximately equal. However, when the microcomponents are added through coprecipitation the size of the particles in the resultant precipitate is generally more or less equal in both the principal component and the microcomponent and it is thus fair to say that no compositional uniformity at the micro level has been achieved. In other words, coprecipitation is basically a fractionating process and the precipitate is a mixture. One of the many techniques which avoids this weakness of the coprecipitation method by creating a uniform blend of component atoms at the atomic level is known as the compound precipitation technique.

5.2.2 Compound Precipitation

In compound precipitation the metallic ions in solution precipitate in the form of a compound with the same stoichiometry as the batch composition. Thus, when the ratio of metallic elements in the precipitated particles matches the ratio of metallic elements in the target compound then the precipitate has compositional uniformity on the atomic level. However, in compounds composed of two or more different metallic elements the ratio of the metallic elements normally forms a simple integral ratio in accordance with what is known as the law of multiple proportion, and it is thus no easy matter to add an appropriate quantity of the microcomponent.

Reasonable uniformity of distribution of the microcomponent at the atomic level in compound precipitation has been achieved by making use of the formation of a solid solution. However, the use of this method is severely restricted in that the system of reaction required to produce a solid solution is itself limited and the composition of the solid solution precipitate generally differs from that of the batch component. If fine particles of the required composition are to be produced then the composition of the solution and of the precipitate must both be rigorously controlled. Figure 5.3 illustrates one type of device used in the synthesizing of compound precipitates using oxalates. Oxalates have frequently been used experimentally for synthesis by means of the compound precipitation technique and compounds such as $BaTiO_3$, $BaSnO_3$ and $CaZrO_3$ have been synthesized from $BaTiO(C_2O_4)_2 \cdot 4H_2O$, $BaSn(C_2O_4)_2 \cdot O \cdot 5H_2O$ and $CaZrO(C_2O_4)_2 \cdot 2H_2O$, respectively. There have also been reports of the experimental synthesis of $LaFeO_3$ from cyanides such as $LaFe(CN)_6 \cdot 5H_2O$. Compound precipitation is capable of producing fine particle precipitates with outstanding compositional uniformity. These fine particles can then be heated to produce the required fine particle compound but there is some debate over whether or not this compositional uniformity of the fine particle precipitate is retained after the heat treatment. In other words, if we look at the synthesis of fine particles of $BaTiO_3$ from $BaTiO(C_2O_4)_2 \cdot 4H_2O$ we find that the calcining of the precipitate $BaTiO(C_2O_4)_2 \cdot 4H_2O$ results in thermal decomposition as shown below [5–7].

$$BaTiO(C_2O_4)_2 \cdot 4H_2O \rightarrow BaTiO(C_2O_4)_2 + 4H_2O$$

$$BaTiO(C_2O_4)_2 + \tfrac{1}{2}O_2 \rightarrow BaCO_3 \text{ (amorphous)}$$
$$+ TiO_2 \text{ (amorphous)} + CO + 2CO_2$$

$$BaCO_3 \text{ (amorphous)} + TiO_2 \text{ (amorphous)} \rightarrow BaCO_3 \text{ (crystalline)}$$
$$+ TiO_2 \text{ (crystalline)}$$

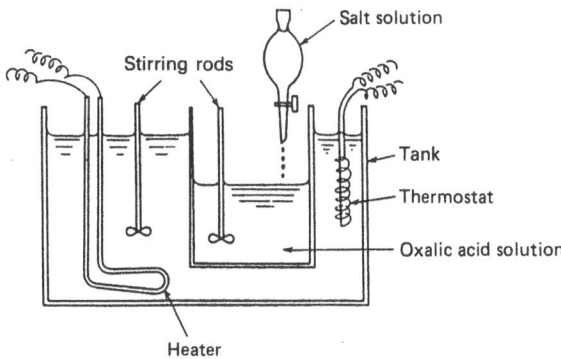

Fig. 5.3 Synthesizing device for compound precipitation using an oxalate.

$BaTiO_3$ is thus not synthesized directly by the thermal decomposition of the fine particle precipitate $BaTiO(C_2O_4)_2 \cdot 4H_2O$ but is formed by virtue of a solid phase reaction which takes place following decomposition of the precipitate into barium carbonate and titanium dioxide. This synthesizing reaction starts to take place at the comparatively low temperature of 450°C due to the highly active condition of the fine particles of barium carbonate and titanium dioxide generated through the process of thermal decomposition. For the creation of full single phase barium titanate, however, temperatures in the region of 750°C are required. At intermediate temperatures there are also a variety of intermediate products which play a part in the formation of the barium titanate. Levels of activity in respect of these intermediate product reactions also vary considerably, which leads to the loss of the outstanding stoichiometric quality originally possessed by the $BaTiO(C_2O_4)_2 \cdot 4H_2O$ precipitate. The synthesis of fine particles by means of compound precipitation almost always takes place via the formation of such intermediate products with the result that the greater the variation between the thermal stability levels of the various intermediate products formed, the greater the lack of compositional uniformity in the fine particles which are eventually produced.

5.3 Hydrolysis

A large number of compounds are known to induce precipitation through the action of hydrolysis and several of them are used as a means of synthesizing superfine particles. Generally speaking the reaction products fall into the categories of hydroxides or hydrates. Since the basic raw materials which form the target for the hydrolytic reaction are metallic salts and water, superfine particulates with a high degree of purity are easily obtained providing the metallic salts themselves are meticulously refined. A number of these chemical compounds, including inorganic salts such as chloride, sulfate, nitrate and ammonium salt, have long been known and studied as colloidal materials in colloid chemistry. In addition to the inorganic salts, metallic alkoxides have also been attracting considerable attention recently as raw materials for use in the synthesis of superfine particles. We have therefore divided the following discussion of hydrolysis into two parts, one relating to the inorganic salts and the other relating to metal alkoxides. Since the metal alkoxides have only recently come to be used for the synthesis of superfine particles our knowledge concerning their characteristics and the mechanism by which they achieve synthesis is still fairly limited and so we have also touched briefly on these points as part of the following section.

5.3.1 Hydrolysis of Inorganic Salts

The synthesis of superfine particles by colloidal techniques using solutions of metallic alum salts, sulfates, chlorides or nitrates has long been known as a way of synthesizing metallic or hydrated oxides. Recently, by controlling the conditions under which hydrolysis takes place, these elements have been used in the synthesis of monodispersed spherical and cubic fine particles and for materials research and the synthesis of new materials. A variety of other applications as, for example, catalyzer, fillers, surface coating agents and photoconductive materials, are also currently under consideration. At present, particular interest is being focused on the possibility of using them as pigments in view of the wide range of tones exhibited by different particle diameters. The synthesis of fine particles of TiO_2 and Fe_2O_3 is introduced in this chapter.

Fine particles of TiO_2 are not in principle difficult to obtain by precipitation using the technique of hydrolysis on a solution of titanate. We shall now introduce a method for the synthesis of spherical fine particles of TiO_2 in monodispersion [8].

The first step in the process involves the distillation at 59°C and 1 mmHg of reagent grade titanium tetrachloride, to which is then added 12 M of hydrochloric acid to form a solution of roughly 1 M in respect of the titanium tetrachloride. At this point the titanium tetrachloride is initially filtered through a 0.22 μm millipore filter to remove any impurities which may have been generated during hydrolysis. The titanium tetrachloride solution to which the hydrochloric acid is added remains highly stable over prolonged periods when subjected to hydrolytic action. When the titanium concentration of the solution has been determined by measurement at an absorbancy level of 400–420 nm of the peroxotitanium which has been generated by the addition of hydrogen peroxide to the solution, then the solution itself is ready for experimental use.

The TiO_2 sol itself, which is formed from particles in monodispersion, is synthesized by the hydrolysis of the titanium tetrachloride solution, which contains ions of sulfuric acid, at a high temperature over a period of anything from a few days to several weeks. For example, the hydrolysis and ageing over a period of 37 days at 98°C of a solution containing concentrations of titanium tetrachloride at 0.106 M and HCl at 5.76 M giving a ratio of $[SO_4^{2-}]/[Ti^{4+}] = 1.9$ will result in the synthesis of fine particles of TiO_2 in monodispersion with a minimum diameter of 2.1 μm, a maximum diameter of 3.1 μm and a central diameter value of 2.6 μm. In this case SO_4^{2-} ions are added in the form of Na_2SO_4. The diameters of the synthesized particles and the

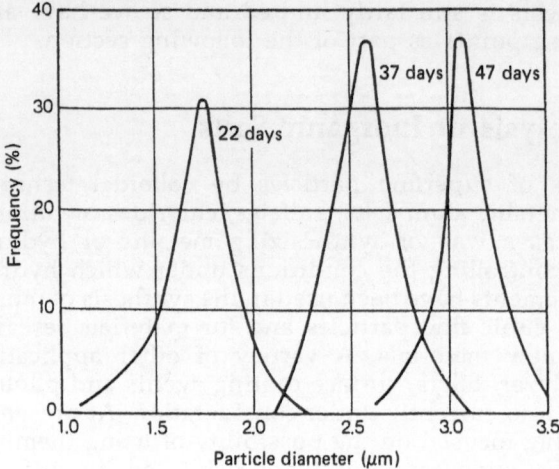

Fig. 5.4 Changes in particle size and size distribution accompanying the ageing of a monodispersion of TiO_2 sol [8]. $TiCl_4$: 0.106 M, HCl: 5.76 M, $[SO_4^{2-}]/[Ti^{4+}] = 1.9$, ageing temperature: 98°C.

corresponding distribution curves depend on ageing time, as illustrated in Fig. 5.4. Basically, particle diameter increases with the length of the ageing time but the induction period required for the growth process to take place makes it impossible to detect any particles in the solution until this period has elapsed. Under the experimental conditions outlined above this period is roughly 15 hours. It has been suggested that the observation of such an induction period in the generation of TiO_2 particles is due to the fact that the particle generation process is essentially a nuclear one or, in other words, that it is related to the supersaturation level of the solution. In the synthesis of colloidal superfine particles by the hydrolysis of inorganic salts it is possible to obtain not only the type of spherical particle just described but also cuboid particles. The following example is of this type and relates more specifically to the synthesis of die shaped superfine particles of hematite [9].

Fine particles of hematite α-Fe_2O_3 are easily synthesized by the hydrolysis of ferric salts. The resultant particles, however, are neither uniform in terms of shape and size, nor do they form a monodispersion. The cuboid shape of the particles is obtained in the following way. Ferric chloride is dissolved in redistilled water at room temperature until a concentrated solution of 2.1–3.6 mol/dm³ is obtained. This solution is then passed through a 0.2 μm millipore filter to obtain the basic hydrolytic solution. Hydrochloric acid is then added to the solution again to bring it back to the required concentration, and it is then mixed with a solution made up of equal parts of water and ethanol in order to produce the sample solution. This sample solution is then introduced into a hermetically sealed container and aged at a high temperature to obtain the required fine particles. The fine particles generated are a mixture of two types: hydrated iron oxide β-FeOOH and hematite α-Fe$_2$O$_3$. The sizes of the two types of particle vary considerably but can easily be separated either by spinning in a centrifuge for 30 minutes at 3000 rpm or by allowing them to undergo a process of natural sedimentation over a period of 1–5 days. β-FeOOH is always rod shaped whereas α-Fe$_2$O$_3$ assumes different shapes depending on the level of concentration of the system of reaction. Figure 5.5 illustrates the relationship between the concentrations of FeCl$_3$ and HCl in the sample solution and particle shape. The area included inside the dotted line indicates the monodispersion of cuboid particles of α-Fe$_2$O$_3$. The pH of the basic sample solution is 1.7 but this ratio falls to 1.0–1.4 following ageing. The angular growth of the cuboid α-Fe$_2$O$_3$ is a function of ageing temperature and time. This relationship is illustrated in Fig. 5.6. When the alcohol was exchanged for propanediol and t-butyl alcohol it was found that with a concentration of 30 vol % or more cuboid α-Fe$_2$O$_3$ could be synthesized, whereas when ethanol was used cuboid α-Fe$_2$O$_3$ could not be synthesized under any conditions.

5.3.2 Alkoxide Hydrolysis

The metal alkoxides [10] are a type of organic metal compound which can be expressed by the general formula M(OR)$_x$ and may be thought of either as derivatives of the alcohol ROH where the hydroxyl group member H has been replaced by the metal M or as derivatives of the metal hydroxide M(OH)$_x$ where hydroxyl group member H has again

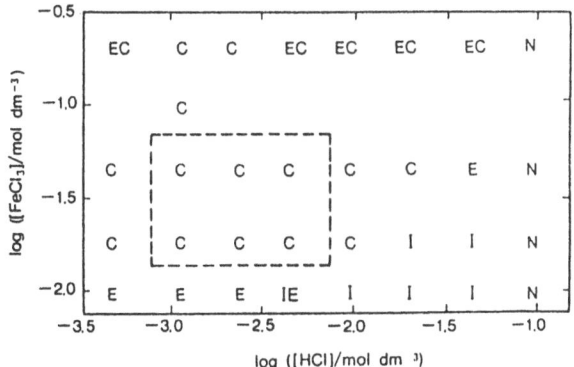

Fig. 5.5 Relationship between shapes of particles generated and concentrations of FeCl$_3$ and HCl in an aqueous 50 vol % ethanol solution [9]. C: cuboid, E: ellipsoid, I: irregular shape, N: no precipitation.

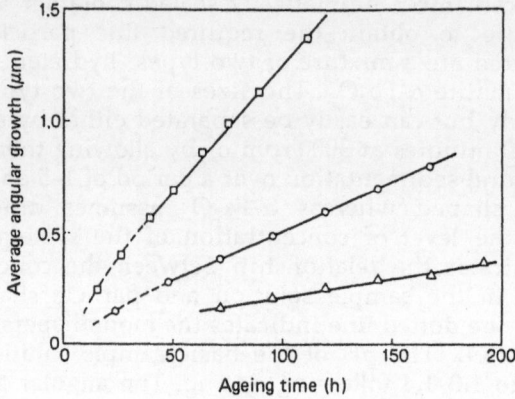

Fig. 5.6 Relationship betwen angular growth and ageing time of cuboid α-Fe$_2$O$_3$, using an aqueous 50 vol % ethanol solution containing an FeCl$_3$ concentration of 1.9×10^{-2} mol/dm^{-3} and an HCl concentration of 1.2×10^{-3} mol/dm^{-3}. $\triangle$: 80°C, $\bigcirc$: 90°C, $\square$: 99°C [9].

been replaced, this time by the alkyl group member R. Historically the alkoxides have been taken to be hydroxy derivatives of metals since they behave either like bases or like oxoacids depending on the electronegativity of the metal elements in question. In other words, they are by convention frequently referred to as alkylorthoesters or arylorthoesters such as orthosilicate, orthoborate and orthotitanate. For example, silicon ethoxide is commonly known as orthoethylsilicate.

The metal alkoxides all have M–O–C bonds with the M–O bond strongly polarized by the strong electronegativity of the oxygen atom which thus gives $M^{\delta+} - O^{\delta-}$. The level of polarization in the alkoxide molecule is related to the electronegativity of the metallic element M. Alkoxides of elements such as sulfur, phosphorus or germanium which carry strong negative charges are actively covalent and behave almost as if they were monomers in terms of volatility. On the other hand the alkoxides of elements such as alkalis, alkaline earth metals and lanthanides which carry strong positive charges associate freely by virtue of their ionic characteristics, thereby exhibiting the characteristics of oligomers. In derivatives of the same metallic element belonging to different alkyl groups the covalence of the M–O bond increases in accordance with the strength of the +I induction effect of the particular alkyl group in question.

The fact that the alkoxides can be thought of as hydroxy derivatives indicates that in terms of chemical characteristics they closely resemble hydroxy compounds. The hydroxy derivatives of the commoner elements range from those with strong base characteristics and a positive charge such as NaOH, Ba(OH)$_2$ and Ln(OH)$_3$ through to acids such as HOCl and (OH)$_3$PO. The basic or acid characteristics of hydroxy derivatives which are obtained by exchanging the alkyl group of the H atom tend to be reduced by the change. However, alkoxides of elements which carry strong positive charges, such as the alkaline metals, also function as extremely strong bases after exchange when in their parent alcohol. Thus, just as it is possible to generate a hydroxo salt such as $Na_2Zn(OH)_4$ by reacting an alkali such as sodium hydroxide with an amphoteric hydroxide such as zinc hydroxide, so it is also possible to synthesize alkoxo salts such as $Na_2[Sn(OEt)_4]$ or $Mg[Al(OEt)_4]$ which correspond to the hydroxo salts of hydroxides by reacting an alkaline alkoxide with a strong base with the alkoxide of an element such as zinc or aluminum.

$$2NaOH \text{ (Alkali)} + Zn(OH)_2 \text{ (Acid)} \rightarrow Na_2Zn(OH)_4 \text{ (Hydroxo salt)}$$

$$2NaOEt \text{ (Alkoxide)} + Zn(OEt)_2 \text{ (Alkoxide)}$$
$$\rightarrow Na_2[Zn(OEt)_4] \text{ (Alkoxo salt)}$$

This sort of neutralization reaction of acid and basic alkoxides depends in the main on the electrical characteristics of the metallic elements of which the alkoxides are composed. The same type of reaction is also known to take place between alkaline earth metals and rare earth metals which carry weaker positive electrical charges than alkaline elements and metal alkoxides such as aluminum, gallium, niobium and tantalum. Some examples of this type of reaction are shown below.

$$M^IOR + M(OR)^{III} \rightarrow M^I[M^{III}(OR)_4]$$

$$M^{II}(OR)_2 + 2M^{III}(OR)_3 \rightarrow M^{II}[M^{III}(OR)_4]_2$$

$$M^{II}(OR)_2 + 4M^{IV}(OR)_4 \rightarrow M^{II}[M_2^{IV}(OR)_9]_2$$

$$M^{II}(OR)_2 + 3M^{IV}(OR)_4 \rightarrow M^{II}[M_3^{IV}(OR)_{14}]$$

$$M^{II}(OR)_2 + 2M^{IV}(OR)_5 \rightarrow M^{II}[M^V(OR)_6]_2$$

(R = alkyl group. M_1 = Li, Na, K, Rb, Cs. M_2 = Mg, Ca, Sr, Ba. M_3 = Al, Ga. M_4 = Zr, Hf. M_5 = Nb, Ta.)

So far, large numbers of metal alkoxides in the periodic table have been synthesized. However, these have all been viewed from the standpoint of organic synthesis. When it comes to fine particle synthesis and material design there is very much less to go on and we have no choice but to await the outcome of future research. Figure 5.7 shows those elements in the periodic table from which alkoxides have so far been synthesized. The sections enclosed by the bold lines are such elements. Research has shown that some of these substances within this line are unsuitable from the viewpoint of fine particle synthesis. Techniques for synthesizing metal alkoxides generally revolve around the electronegativity of the metal elements.

Elements with strong positive electrical charges represented by

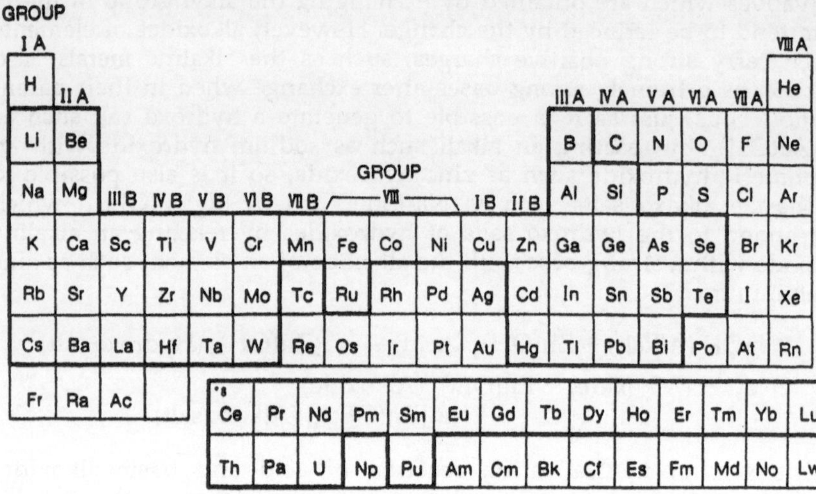

GROUP

I A																	VIII A
H	II A											III A	IV A	V A	VI A	VII A	He
Li	Be											B	C	N	O	F	Ne
Na	Mg	III B	IV B	V B	VI B	VII B		— VIII —		I B	II B	Al	Si	P	S	Cl	Ar
K	Ca	Sc	Ti	V	Cr	Mn	Fe	Co	Ni	Cu	Zn	Ga	Ge	As	Se	Br	Kr
Rb	Sr	Y	Zr	Nb	Mo	Tc	Ru	Rh	Pd	Ag	Cd	In	Sn	Sb	Te	I	Xe
Cs	Ba	La	Hf	Ta	W	Re	Os	Ir	Pt	Au	Hg	Tl	Pb	Bi	Po	At	Rn
Fr	Ra	Ac															

Ce	Pr	Nd	Pm	Sm	Eu	Gd	Tb	Dy	Ho	Er	Tm	Yb	Lu
Th	Pa	U	Np	Pu	Am	Cm	Bk	Cf	Es	Fm	Md	No	Lw

Fig. 5.7 The metallic elements so far reported to have been synthesized to form alkoxides (within the bold lines).

valence numbers up to 3, such as the alkaline metals, alkaline earth metals and lanthanides can be synthesized in direct reaction with alcohol

$$M + nROH \rightarrow M(OR)_n + \frac{n}{2}H_2$$

However, with metals such as magnesium, beryllium, aluminum and the lanthanides which carry only weak positive charges, it is necessary to introduce a catalyst such as mercuric chloride to help the reaction along. The precise role of the catalyst in this reaction is still not clearly understood but it is thought likely that the answer lies either in the simple cleansing of the surface of the metal or else in the generation of an intermediate derivative such as a chloride, which facilitates reaction with the alcohol. The ease with which the reaction between the metal and the alcohol progresses increases along with the positive charge of the metal but it is also influenced by the characteristics of the alcohol itself in that a reaction with the same metal slows with the increased branching of the alcohol. This is attributable to the reduction of the acidity of the side-chain alcohol resulting from the alkyl group's +I induction effect.

In the case of alkoxides which cannot be synthesized by means of a direct reaction between a metal and an alcohol it is customary to use one of the metal halides, normally a chloride, in place of the original metal. The kind of reaction which takes place between a chloride and an alcohol is known as a nucleophilic substitution reaction between two molecules. The ease with which the chlorine ion and the alkoxide anion change places in this reaction is heavily influenced by the level of electronegativity of the metallic element of the chloride which is subjected to the nucleophilic attack. For example, silicon, titanium, zirconium and thorium each has a lower level of electronegativity than the previous element in the list (silicon having the highest level). The chloride of each of these elements also exhibits a correspondingly lower level of reactivity with alcohol until eventually the complete substitution of chlorine ion with the alkoxide group fails to take place. Complete substitution can, however, be achieved with the help of a base such as ammonia, pyridine or trialkylamine. This is because it increases the concentration of alkoxide anions in the system by virtue of the type of reaction described below:

$$B + ROH \rightarrow (BH)^+ + (OR)^-$$

$$(OR)^- + MCl \rightarrow MOR + Cl^-$$

$$(BH)^+ + Cl^- \rightarrow (BH)Cl$$

The metal alkoxides react with the water to induce the precipitation of oxides, hydroxides and hydrates. In most cases the reaction is extremely fast, although there are alkoxides such as those of silicon and phosphorus which only react slowly. Drying in the case of an oxide precipitate or calcining in the case of hydroxides or hydrates will convert such precipitates into fine ceramic powders. A number of compound oxides useful in the production of ceramics have so far been synthesized, as shown in Table 5.2. The following example, which

Table 5.2 Oxide powders synthesized from alkoxides classified according to the precipitation conditions

Crystalline powder	$BaTiO_3$, $SrTiO_3$, $BaZrO_3$, $Ba(Ti_{1-x}Xr_x)O_3$, $Sr(Ti_{1-x}Zr_x)O_3$, $(Ba_{1-x}Sr_x)TiO_3$, $MnFe_2O_4$, $CoFe_2O_4$, $NiFe_2O_4$, $ZnFe_2O_4$, $(Mn_{1-x}Zn_x)Fe_2O_4$, Zn_2GeO_4, $PbWO_4$, $SrAs_2O_4$
Crystalline hydroxide powder (oxidized by calcination)	$BaSnO_3$, $SrSnO_3$, $PbSnO_3$, $CaSnO_3$, $MgSnO_3$, $SrGeO_3$, $PbGeO_3$, $SrTeO_3$
Amorphous powder (oxidized by calcination without passing through an intermediate phase)	$Pb(Ti_{1-x}Zr_x)O_3$, $Pb_{1-x}La_x (Zr_{1-y}Ti_y)1-_{x14}O_3$, $Sr(Zn_{1/3}Nb_{2/3})O_3$, $Ba(Zn_{1/3}Nb_{2/3})O_3$, $Sr(Zn_{1/3}Ta_{2/3})O_3$, $Ba(Zn_{1/3}Ta_{2/3})O_3$, $Sr(Fe_{1/2}Sb_{1/2})O_3$, $Ba(Fe_{1/2}Sb_{1/2})O_3$, $Sr(Co_{1/3}Sb_{2/3})O_3$, $Ba(Co_{1/3}Sb_{2/3})O_3$, $Sr(Ni_{1/3}Sb_{2/3})O_3$, $NiFe_2O_4$, $CuFe_2O_4$, $MgFe_2O_4(Ni_{1-x}Zn_x)Fe_2O_4$, $(Co_{1-x}Zn_x)Fe_2O_4$, $BaFe_{12}O_{19}$, $SrFe_{12}O_{19}$, $PbFe_{12}O_{19}$, $R_3Fe_5O_{12}$(R=Sm, Gd, Y, Eu, Tb), $Tb_3Al_5O_{12}$, $R_3Gd_5O_{12}$(R=Sm, Ge, Y, Er), $RFeO_3$(R=Sm, Y, La, Nd, Gd, Tb), $LaAlO_3$, $NdAlO_3$, $R_4Al_2O_9$(R=Sm, Eu, Gd, Tb), $Co_3As_2O_8$, $(Ba_xSr_{1-x})Nb_2O_6$

relates to the production of $BaTiO_3$, will serve to illustrate this synthesizing process [10–14].

The basic raw materials required for the synthesis of $BaTiO_3$ are the alkoxides of Ba and Ti. Ba alkoxide is formed by reacting metallic Ba directly with alcohol while Ti alkoxide is formed by reacting titanium tetrachloride with alcohol while blowing NH_3 gas. When the reaction has finished the solvent is replaced with benzene and the secondary product NH_4Cl filtered out. When the concentrations of the Ba and the Ti have been determined the two metal alkoxides are mixed together in solution until a ratio of Ba:Ti = 1:1 is achieved and then refluxed again for another 2 hours or so. The solution is agitated continuously while distilled water is added gradually to hydrolyze it. The hydrolytic reaction will result in the precipitation of a superfine, white, crystalline particulate of $BaTiO_3$. Figure 5.8 is a flow diagram illustrating this synthesizing reaction.

Whichever alcohol is used in this reaction, be it methanol, ethanol, *i*-propanol, *n*-butanol or whatever, the powder which is ultimately synthesized will have essentially the same sort of characteristics. In other words, the effect of the alkoxide's alkyl group on the size and shape of the particles in the powder is negligible. Whichever one is used, single phase crystalline barium titanate is still obtained. Little difference was found between the X-ray diffraction patterns of the particles of the various barium titanates obtained. Close resemblances were also found between particles subjected to differential thermal analysis or observed through an electron microscope. Figure 5.9 shows the X-ray charts of barium titanates synthesized from a variety of different alcohols. The higher the boiling point of the particular alcohol used, the better the crystallinity of the powder was found to be under X-ray but, on the other hand, little difference was observed in terms of particle size, shape or aggregation conditions when observed through an electron microscope. This type of particulate oxide which is generated by the hydrolysis of metal alkoxides is virtually insoluble in the reaction solution, which in

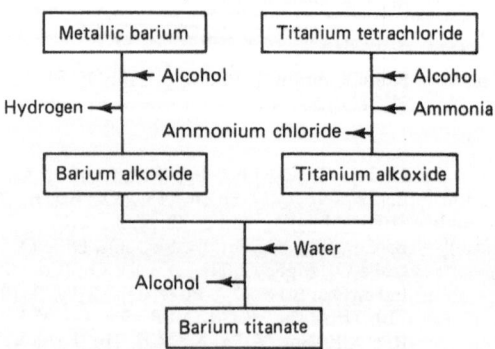

Fig. 5.8 Flow diagram illustrating the synthesis of barium titanate.

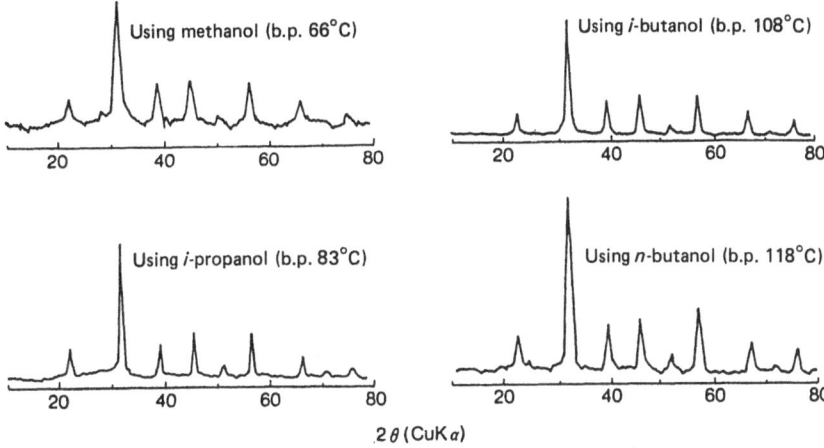

Fig. 5.9 X-ray charts of barium titanate powder synthesized from various alkoxides.

turn means that particle growth as a result of ageing after synthesis cannot be expected. In other words, the diameters of the synthesized particles must be determined at the hydrolytic reaction stage. Under experimental conditions the most effective variable in the control of particle size was found to be the level of concentration of the alkoxide used in the hydrolytic system of reaction. Figure 5.10 illustrates the relationship between alkoxide concentration and particle diameter observed during the synthesis of barium titanate using alkoxide concentrations ranging from approximately 0.01 to 1 mol/l. The graph clearly

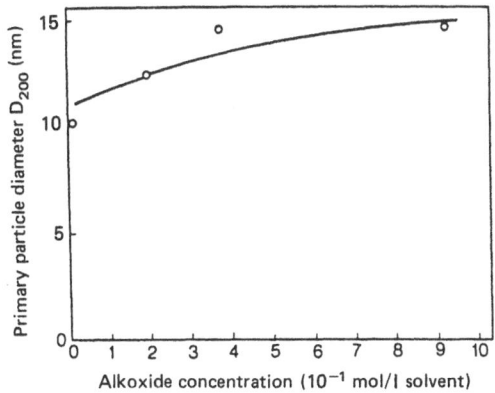

Fig. 5.10 Relationship between the size of crystallites of superfine $BaTiO_3$ particles and the alkoxide concentration in a reaction system [13].

demonstrates the insignificance of the change in size of the primary particles of BaTiO$_3$ generated even in response to a hundredfold increase in the concentration level of alkoxide in the reaction system. Particle size in fact varied from 10 nm at low concentration levels up towards 15 nm as the concentration level was increased. Generally speaking, therefore, the sizes of particles in powders synthesized from alkoxides are little affected by the nature of the alkoxide, and normally range between 10 and 100 nm. Changes in the variables of the reaction system have virtually no effect on the sizes of the particles synthesized.

In the synthesis of fine particles from the metal alkoxides it is quite common, as was seen in the case of BaTiO$_3$, to obtain fine oxide particles in a crystalline form directly via the hydrolytic reaction. Typical examples of the BaTiO$_3$ type are SrTiO$_3$, BaZrO$_3$, CoFe$_2$O$_4$, NiFe$_2$O$_4$ and MnFe$_2$O$_4$ or solid solutions such as (Ba, Sr)TiO$_3$, Sr(Ti, Zr)O$_3$ and (Mn, Zn)Fe$_2$O$_4$. Since all these fine particles are crystalline in structure their composition necessarily matches the batch composition of the system of reaction at the particle level. Table 5.3 records the results of a compositional analysis of SrTiO$_3$ particles synthesized from a metal alkoxide. The analysis itself was carried out on 50 particles using an energy dispersion analyzer under a transmission scanning electron microscope. The ratio of Sr to Ti in the SrTiO$_3$ particles at each of the different alkoxide concentrations was very close to unity and the particles themselves also exhibited a high degree of compositional uniformity. The results of these experiments again showed a tendency for the compositional variation between particles to increase along with increases in the alkoxide concentration but this was attributable in part to a certain lack of uniformity in the alkoxide solutions, in that the two solutions containing the lower concentrations of alkoxides were completely transparent while the solution containing the high concentration was cloudy. The important point to take note of here is not that the compositional variation between particles was greater at higher concentrations but that there was in fact still an extremely high level of uniformity achieved despite hydrolysis taking place in a reaction

Table 5.3 Compositional analysis by electron microscope of fine particles of SrTiO$_3$ synthesized from alkoxides [13]

Alkoxide concentration (solvent: mol/l)	Water volume for hydrolysis (multiples of theoretical volume)	Reflux period following hydrolysis (h)	Anion ratio			
			Average		Standard deviation	
			Sr	Ti	Sr	Ti
0.117	20×	4	1.005	0.998	0.030 2	0.015 1
0.616	20×	2	1.009	0.996	0.045 8	0.022 8
3.61	6.8×	2	1.018	0.991	0.062 9	0.031 4

system where the solution containing the solid phase suspension was considerably lacking in uniformity.

Generally speaking, the more basic the elements of which a particle is composed, the more amorphous will be the fine particles obtained. The composition of such amorphous particles is nevertheless batch composition at the particle level. It is of course obvious that systematic uniformity of composition is unlikely to be achieved in solid phase reactions but this also turns out to be the case with gas and liquid phase reactions which would at first glance appear to offer the possibility of complete mixing at the atomic level. This is because there is always a certain amount of time required for the fine particle generation system to reach the requisite physical and chemical conditions. The shorter the time required to attain the appropriate environmental conditions, the closer the elements in the system will approach a state of total mixture. The hydrolysis of all metal alkoxides, however, takes place at a much faster rate than the time taken for the metallic elements in the solution to lose their uniformity. It seems likely therefore that the mixture of metallic elements in the precipitate closely reflects the initial levels of mixing within the solution, thereby giving rise to an amorphous particle structure. The outstanding feature of the fine particles generated from alkoxides resides in their ability to reflect precisely at fine particle level the batch composition of the synthesizing reaction. This in turn makes possible a hitherto unattainable degree of precision in the design of a fine particle's structure during subsequent heat treatment.

Generally speaking, heat treatment has the effect of reducing a substance which is mixed at the atomic level to the most stable set of phase relationships for the substance in question. Moreover, even in systems where a total mix has been achieved at the atomic level the phase changes resulting from heat treatment cannot be defined simply in terms of the basic thermodynamic relationships existing between members of the system. Even in a simple system consisting of just two members, there nevertheless exist a large number of complex phase relationships such that, while there may be particle growth taking place on a higher systematic level, a look inside each particle may in fact reveal a process of chemical separation and reconstitution taking place. The most important feature of the changes introduced into a system during heat treatment, however, is not their level of complexity but the fact that they are generated *in situ*. In other words, the structure of fine particles synthesized from alkoxides can be controlled *in situ* during heat treatment. On the other hand, the generation of fine particles by solid phase reaction is designed for the introduction of uniformity to a system which initially lacks uniformity. It is thus extremely difficult to synthesize a uniform powder where all the particles have an identical composition by means of a solid phase reaction.

Research into the synthesis of powders from alkoxides has thus been concentrated largely on solid solution systems. Alkoxides have, however, been used successfully for the formation of fine particles *in situ* within solids, thereby providing us with access to a new materials production technique making use of the surfaces which are generated within a solid by the destruction of its uniformity. This is an extremely attractive development from the point of view of the expansion of the range of functions of ceramics.

5.4 Atomization

The most commonly used physical technique for the synthesis of fine particles using a solution as the basic raw material is probably atomization. This techniqiue can itself be broken down into several different types depending on the way in which the drops of liquid are dealt with after undergoing gas phase atomization. The droplets may simply be dried and collected as they are or alternatively they may be subjected to heat treatment in order to produce the required compound particles, using what is termed the atomized drying technique. Other techniques include the atomized hydrolysis technique whereby the droplets are subjected to gas phase hydrolysis, or the atomized calcination technique whereby the droplets are first dried whilst suspended in the gas phase and subsequently subjected to heat treatment.

5.4.1 Atomized Drying

The atomized thermal decomposition technique involves the firing of the atomized solution or slurry of raw materials through a nozzle in order to produce the required fine particles. Figure 5.11 shows a schematic representation of a device used to synthesize superfine particles of soft ferrite. The basic process requires the introduction of the metal salt in the form of a solution into the atomizer where atomization takes place. The atomized and dried salt is then collected in a cyclone where it is heated by a furnace to generate the required fine particles. The basic raw material consists of an aqueous solution of the sulfates of nickel, zinc and iron which is atomized to produce spherical particles of mixed sulfates some 10–20 μm in diameter [15]. If these particles are then calcined at 800–1000°C nickel–zinc ferrite will be obtained. The powder thus obtained by calcination consists of an aggregation of primary particles about 0.2 μm in size which can easily be mixed in a turbine mixer to produce a submicron size superfine particulate. The use of special microwaves at high output levels requires a strong critical magnetic field to excite spin waves which in turn necessitates particularly small particles of ferrite. However, the strengthening of the critical magnetic field always has the effect of reducing the electroconductivity of the material. It is known that this reduction is due to the lack of

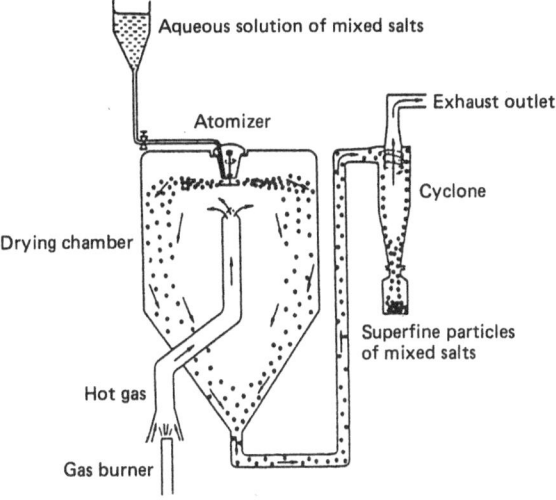

Fig. 5.11 Schematic representation of atomized dryer [15].

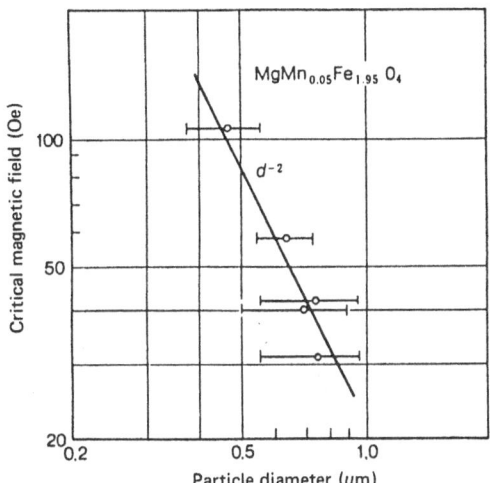

Fig. 5.12 Relationship between particle size and strength of critical magnetic field of Mg–Mn ferrite obtained by hot pressing atomized, dried powder [15].

uniformity in the material. Figure 5.12 illustrates the relationship between the critical magnetic field required to excite spin waves in magnesium–mangan ferrite obtained by the same process using the same experimental apparatus, and the particle size of the sintered compact. The results clearly indicate the inverse relationship which exists between the size of particle and the strength of the critical magnetic field. The use of this technique involves only a small dielectric loss which means not only that the resultant superfine particles are particularly small but also that their composition is relatively uniform.

5.4.2 Atomized Hydrolysis

The technique for synthesizing fine particles by the hydrolysis of alkoxides was introduced in Sect. 5.3.2. Attempts have also been made to synthesize fine particles in monodispersion by means of a combination of physical and chemical methods, using a physical method such as the formation of an aerosol and a chemical method such as the hydrolysis of the alkoxides. Figure 5.13 shows the type of apparatus used in the synthesis of alumina spheroids by this method [16]. The synthesis itself

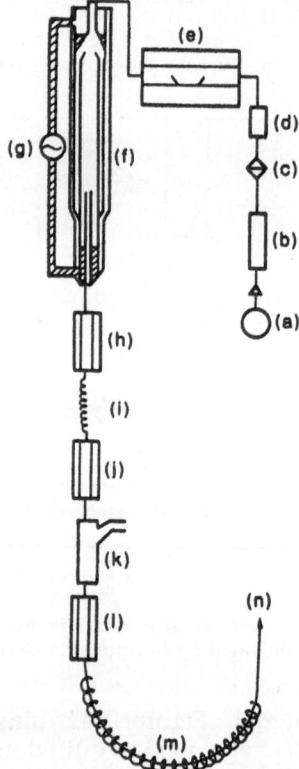

Fig. 5.13 Apparatus for the production of alumina spheroids by atomized hydrolysis [16]. (a) Carrier gas, (b) drying column, (c) filter, (d) flowmeter, (e) atomizing furnace, (f) boiler, (g) pump, (h) condenser, (i) heating area, (j) condenser, (k) hydrolyzer, (l) condenser, (m) heating area, (n) aerosol outlet.

was achieved in the following way. Nuclei of silver chloride are entrained by a carrier gas into an atmosphere of vaporized aluminum butoxide and then cooled. This operation results in the generation of an aluminum butoxide aerosol formed by combination with the vaporized silver chloride. This aerosol is composed of droplets in monodispersion which are then brought into contact with steam to produce a hydrolytic reaction. This in turn produces particles of aluminum hydroxide in monodispersion. This is then calcined to produce particles of alumina.

The carrier gas helium (a) is dried in a magnesium perchlorate and calcium sulfate column (b), filtered through a millipore filter (c) and then passed through an electric furnace (e) where it picks up the nuclei of silver chloride. A tundish containing the silver chloride is positioned in the middle of the electric furnace, which is kept at a constant temperature between 595°C and 650°C. The helium gas which emerges from the electric furnace then passes through a boiler (f) generating aluminum butoxide vapor with which it is saturated in transit. The speed of the gas flow is between 500 and 2000 cm³/min, the boiler temperature between 122°C and 155°C and the alkoxide vapor pressure a little under 1 Torr. The helium gas saturated with alkoxide vapor is then cooled in a condenser (h) to produce an aerosol. The alkoxide in the gas which has not turned to aerosol is reheated up to around 130°C in a heating area (i) to revaporize it completely and is then passed into a condenser (j) which is kept at 25°C to condense it again. This has the effect of turning the remaining alkoxide in the helium carrier gas to a pure alkoxide liquid aerosol. The aerosol is then mixed with water vapor in the hydrolyzer (k) and passed through a condenser (l) held at 25°C to ensure complete hydrolyzation. The hydrolyzed aerosol emerges into a glass tube heated to 300°C where it solidifies and is eventually collected in a millipore filter.

Hydrolyzing apparatus comes in many shapes and forms and the type which is represented schematically in Fig. 5.14 is known as the "falling film" type. The apparatus derives its name from the way in which the water film is continually seen to fall on the surface of a glass tube (d) which is held at a constant temperature. The water is circulated constantly with the help of a pump (e) between an upper (b) and lower (c) receptacle. The aerosol is hydrolyzed as it passes through the tube from inlet (a) to outlet (f).

Figure 5.15 shows the size distribution of aluminum butoxide droplets obtained using helium carrier gas at a flow speed of 1000 ml/min, a silver chloride vaporization furnace temperature of 610°C, a boiler temperature of 122°C and a secondary heater temperature of 130°C, plus the size distribution of the aluminum hydroxide obtained by the hydrolytic reaction. The size dispersion of the droplets was determined by measuring the light scatter, and the size dispersion of the aluminum hydroxide after hydrolyzation was determined by means of electron microphotographs. The figure itself clearly shows the solid particles created by the hydrolytic reaction to be smaller than the immediate precursor droplets. The mode

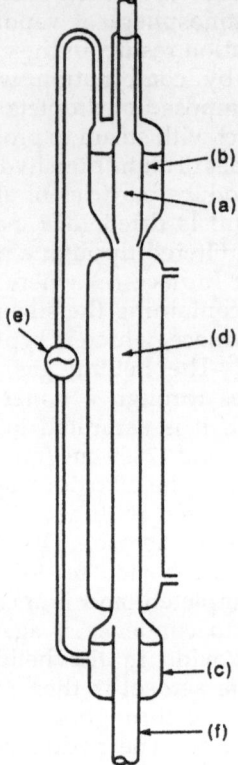

Fig. 5.14 Alkoxide hydrolyzation apparatus of the falling water film type [16]. (a) Inlet for liquid aerosols, (b) upper water receptacle, (c) lower water receptacle, (d) falling film area, (e) pump, (f) outlet for hydrolyzed aerosol.

diameter of the aluminum butoxide droplets is 0.42 μm. The mode size of the compound generated by hydrolytic reaction under experimental conditions is 0.17–0.18 μm but if the generated compound were bermite, for example, or diaspore, then their respective densities of 3.01 g/cm³ and 3.44 g/cm³ would give rise to post-hydrolysis mode sizes of 0.186 μm and 0.178 μm, both of which would be entirely consistent with the same pre-hydrolysis mode size of 0.42 μm. The particles generated by the hydrolysis of the aluminum alkoxide may be thought of as a 1-hydrate of bermite or of diaspore.

5.4.3 Atomizing and Calcining

Figure 5.16 shows a schematic representation of apparatus used for atomizing and calcining [17]. The initial solution is forced through a

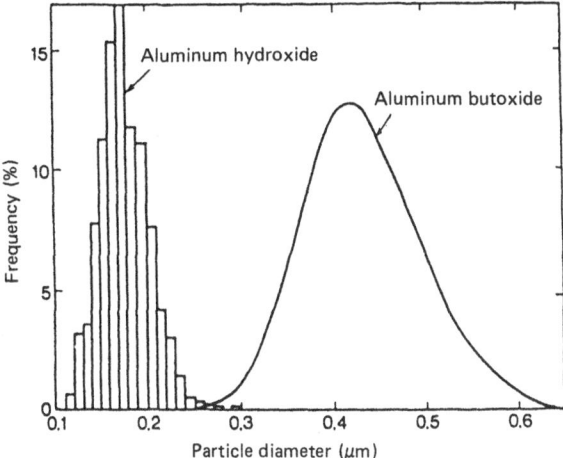

Fig. 5.15 Size dispersions of aluminum butoxide droplets and particles of aluminum hydroxide following hydrolysis [16].

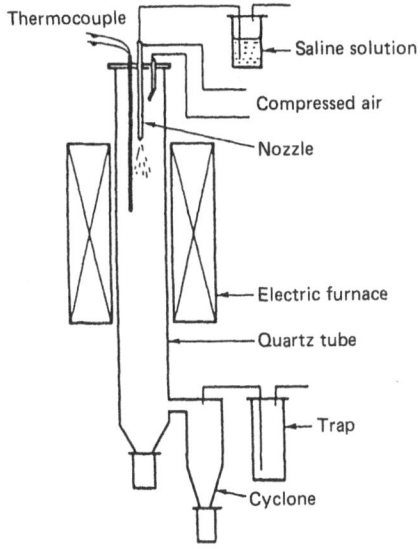

Fig. 5.16 Schematic representation of apparatus for atomizing and calcining [17].

nozzle with compressed air thereby causing the two to mix and atomize. The size of the resulting droplets can be controlled by changing the diameter of the nozzle. The droplets are entrained in the downward flow of air and pass through a quartz tube heated from the outside, where they undergo a process of thermal decomposition which comminutes them into fine particles. This atomization and calcination of the mixed solution of magnesium nitrate and aluminum nitrate results in the synthesis of magnesium–aluminum spinels. The solvent used is a mixture of water and methanol and the size of the particles of the resulting spinels is known to be larger the lower the concentration of salt in the solution or the higher the concentration of methanol in the solvent. Figure 5.17 shows the size distribution of the fine particles generated with a solution supply nozzle diameter of 0.4 mm, a solution flow volume of 4 ml/min, an atomizing air pressure of 1 atmosphere, a flow of 10 l/min and a calcination temperature of 800°C. The particles obtained are composed of single phase spinel consisting of primary particles of several tens of nanometers in diameter. The atomization and calcination method of superfine particle production is not a difficult technique to use and while powder production costs would depend very much on the scale of production they have nevertheless been provisionally estimated at somewhere in the region of several hundreds of dollars per ton.

5.5 Oxidation–Reduction Method

The oxidation–reduction method involves the synthesis of powders consisting of fine particles of metals or oxides by direct oxidation and reduction of the initial substance. The method is used industrially for the production of metallic silicon through the reduction of silicon tetrachloride using zinc vapor, and for the production of zinc oxide and

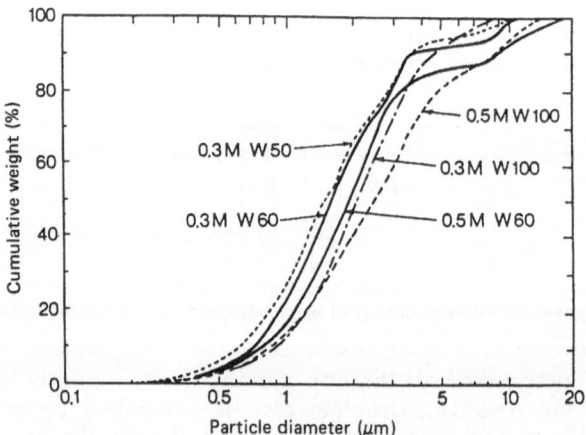

Fig. 5.17 Particle dispersion in spinel powder obtained by the atomization and calcination technique [17].

antimony oxide by the oxidation of zinc and antimony vapor using hot air. Synthesis is achieved in each of the above cases by means of a gas phase reaction although it is also possible to synthesize fine particles by direct oxidation of metals in a liquid or near-liquid phase.

5.5.1 Oxidation in Liquid Phase

The best known examples of the synthesis of particles by oxidation in liquid phase are those relating to the oxides of spinels such as magnetite, manganese ferrite and cobalt ferrite. By inducing the precipitation of ions in the form of fine particles of bivalent hydroxides from a solution of bivalent iron, manganese or cobalt it is possible to oxidize the precipitate while the solution remains in a cloudy state, thereby achieving the synthesis of superfine oxide particles. The synthesis of Fe_3O_4, for example, involves first dissolving iron(II) sulfate $FeSO_4 \cdot 7H_2O$ in oxygen-free water which is being heated and cooled in a current of nitrogen gas. A solution concentration in the region of 10% is about right. Twenty per cent ammonia water is then added to the solution to induce the precipitation of iron(II) hydroxide, thereby making the solution cloudy. If this cloudy solution is then gently oxidized while heating it to something in excess of 70°C, regular superfine octahedral or cuboid particles with sides of uniform lengths in the region of 0.2 μm will be obtained. Either air or an oxidizing agent such as KNO_3 can be used to induce oxidation. The iron(II) hydroxide precipitated varies according to factors such as the type of alkali used for the precipitation process, the volume of precipitation and the oxidation temperature. Figure 5.18 illustrates the relationship between $R = 2NaOH/FeSO_4$, the volume of $Fe(OH)_2$, the oxidation temperature and the generated substance where NaOH is used as the alkali and air is used as the oxidizing agent [8,19].

5.5.2 Hydrothermal Oxidation

Substances which do not oxidize easily in solutions held at a normal temperature and pressure can often be caused to oxidize much more rapidly at high temperatures and pressures. For example, the oxidation of metallic iron due to moisture in the atmosphere takes place extremely slowly. The reaction time can, however, be reduced drastically simply under the hydrothermal conditions. For example, the oxidation process takes just 1 hour under hydrothermal conditions of 98 MPa and 400°C producing a thickness of magnetite ranging from several tens up to 100 micrometers. Even substances such as metallic zirconium, which does not oxidize under normal conditions, can be caused to oxidize under different hydrothermal conditions. Figure 5.19 is an illustration of the type of apparatus and system of reaction used to synthesize a fine powder of Al_2O_3/ZrO_2 from an alloy of Al and Zr. When zirconium powder only

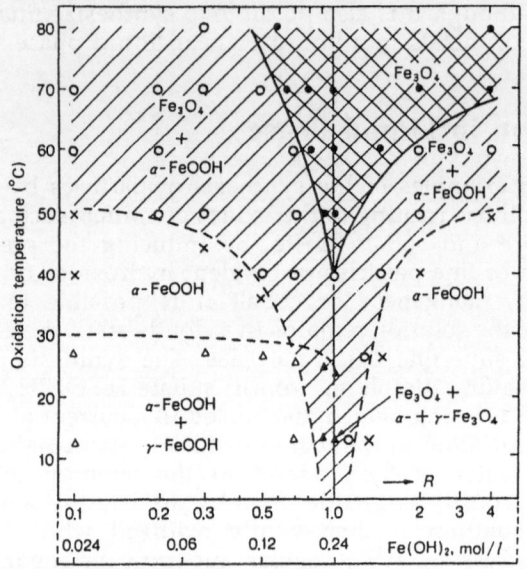

Fig. 5.18 Relationship between the oxide generated from $Fe(OH)_2$ and $R=2NaOH/FeSO_4$, the volume of $Fe(OH)_2$ and the oxidation temperature.

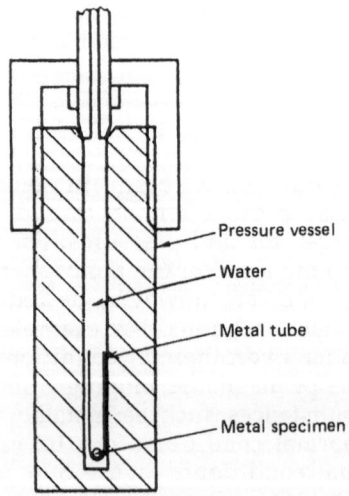

Fig. 5.19 Diagrammatic arrangement of hydrothermal oxidation apparatus.

is subjected to hydrothermal oxidation, monoclinic zirconia in single phase can be obtained at 100 MPa and a temperature between 250°C and 700°C. The diameters of the particles forming the powder are in the region of 25 nm. If aluminum alone is subjected to a similar process at 100 MPa and a temperature of 500°C, γ-AlOOH is generated, while at temperatures between 500°C and 700°C α-Al$_2$O$_3$ is generated. The average particle size of the Al$_2$O$_3$ is between 110 nm and 200 nm. If these two end members are alloyed, however, then the temperature required for complete oxidation to take place varies according to the composition ratios of the two metals in the alloy. For example, an alloy composition of Zr$_5$Al$_3$ requires a temperature between 500°C and 700°C at 100 MPa for complete oxidation to take place. The resultant fine compound powder is composed of monoclinic zirconia, tetragonal zirconia and α-Al$_2$O$_3$ particles between 10 nm and 35 nm in size.

5.6 Freeze Drying

Some 15 years or so have now elapsed since the freeze drying method of producing highly active, highly reactive fine particles was first propounded [4,21]. The great value of this technique rests in the fact that it is at once easy to use, and at the same time offers the possibility of large scale production at relatively low cost. The principal features of this process are as follows:

1. Precise amounts of initial powders with complex compositions can be controlled by utilizing uniform solutions of soluble salts.
2. The uniform mixing of metallic ions in solution can be maintained by rapid freezing.
3. The freeze drying operation also permits the easy control of anhydrous salts. The hydrates of anhydrous salts can be fused at much lower temperatures than those required to fuse the anhydrous salts themselves, which means that the compositional separation of mixed salts as a result of fusion can be avoided.
4. Freeze drying leaves the dried body in a porous condition. As a result, gas permeability is high which means not only that gas generated during calcination can be expelled without difficulty but also that the dried body can easily be comminuted, making it easy to reduce it to a fine particulate.

The freeze drying technique itself involves first creating the correct compositional balance within the solution containing the metallic ions of the target material, and then atomizing the solution to reduce it to fine droplets which are immediately freeze dried and solidified at very high speed. The water is in this way forced out of the frozen droplets and vaporized to produce anhydrous salt. The calcining of this salt at a low temperature has the effect of turning it into a fine powder. Let us now consider a system consisting of two members, salt and water. Since water is commonly used as a solvent we have illustrated in Fig. 5.20 the appropriate temperatures and pressures for a freeze drying process using water as the solvent. At point E four phases – ice, salt, solution and vapor – exist simultaneously. At this point the freedom value of the system is given by the phase rule $F = n+2-P = 2+2-4 = 0$. The figure shows four curves emanating from point E: ice + solution + gas phase, ice + salt + solution, ice + salt + gas phase, and salt + solution + gas phase. Each of the curves on the temperature/pressure graph represents three distinct phases and each therefore has a freedom value of 1. The areas enclosed by these four curves each represent two phases and each has a freedom value of two. Since aqueous solutions are normally adjusted at room temperature and normal atmospheric pressure the start point has been marked on the temperature/pressure graph in the liquid phase area of water, which is the area demarcated by the curves representing the melting and vaporization points of water. We have called this point 1. In this condition the vaporization pressure of the solution can be thought of as equal to the vaporization pressure for pure water. This is because the vaporization pressure of salt in solution at room temperature is small enough to be ignored. If the solution at point 1 is then rapidly chilled then it moves to point 2 on the graph and the system assumes the condition of a solid compound of ice and salt. If the compound is

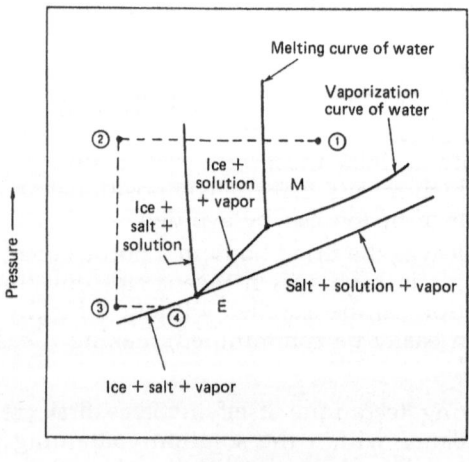

Fig. 5.20 Temperature and pressure graph of a solution of salt and water. M: triple point of water, E: simultaneous four phase point (ice + salt + solution + vapor).

then subjected to a pressure below that of the system's invariant point E it will slowly heat up, thereby moving the system into the salt + vapor area. In other words, the system has been moved from point 1 through points 2, 3 and 4 on the temperature/pressure graph. If, with the system at point 4, the vapor phase is then expelled this will leave only the salt.

During the freezing process itself it is better to reduce the droplets to their smallest possible size in order to prevent separation of the salts held in solution. Such separation of the salts is prevented since the smaller the droplets are the sooner the freezing process can be completed. The frozen liquid droplets are produced by spraying salt solution into the cooling medium. In this case it is necessary for the liquid and the cooling medium to be immiscible. For example, if a salt and water solution is sprayed into liquid hexane which is in the process of being cooled by a freezing mixture of dry ice and acetone it is quite easy to produce ice droplets in the neighborhood of 0.1–0.5 mm in size. The type of laboratory apparatus in common use for this process is shown in Fig. 5.21. It is also possible to spray a salt solution directly into a liquefied gas, such as liquid nitrogen, in place of the hexane which was cooled with a freezing mixture. The hexane temperature of −77°C achieved with the dry ice and acetone freezing mixture may be compared with a temperature of −196°C which can be obtained by direct cooling with liquid nitrogen. From the point of view of temperature, therefore, liquid nitrogen is clearly the more effective freezing agent, but in practice good results are also commonly achieved with hexane. The reason for this is that when liquid nitrogen is used there is always a trace of nitrogen in gas phase around the droplets which impedes the transfer of heat to the surrounding environment. The drying process necessitates the sublimation of the ice phase without melting the frozen droplets. By applying heat at the point of sublimation it is possible to accelerate the drying process. This is because thermal energy is required for sublimation

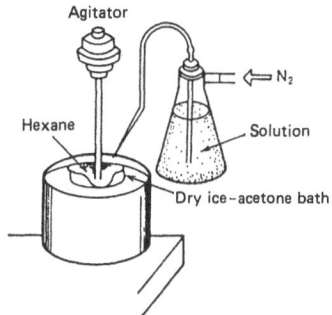

Fig. 5.21 Experimental droplet freezing device.

to take place. When heating causes sublimation of water in the ice droplets, the temperature of the system is maintained at a constant level because the heat of sublimation of the ice depends upon the balance that exists between the self-cooling of the ice droplets and heat supplied from the atmosphere. As the level of vacuum is raised the deterioration in thermal conductivity increases and without an additional supply of heat the drying of the ice droplets becomes a very slow process. In order to assist the drying process by keeping the water vapor pressure of the system as low as possible a condenser is required to collect the sublimated water. Under laboratory conditions a liquid nitrogen trap is used. Figure 5.22 shows a schematic representation of this type of apparatus.

The solubility of a salt has a very significant impact on the effectiveness of the freeze drying process. Generally speaking, the effective collection of water sublimated from frozen droplets in a condenser or cold trap requires a frozen droplet temperature in the region of −10°C. For this reason the salt concentration of the solution must not be too great. On the other hand, however, the processing capacity of the apparatus depends on increasing the solute concentration, which in turn reduces the freezing point of the solution and impairs the efficiency of the apparatus. Moreover, a highly concentrated solution also creates a supercooled condition, reducing the droplets to a vitreous state and causing the salt to separate and the particles to aggregate. One effective way of preventing this is by the addition of ammonium hydroxide to the saline solution. Freeze drying has also been tried with, for example, alumina [22], aluminum–magnesium spinel [23], lithium ferrite [24] and nickel oxide [25]. We will now explain this process in relation to the synthesis of alumina [22].

Aluminum sulfate, $Al_2(SO_4)_3 \cdot 16-18H_2O$, is dissolved in water and brought to a solution of 0.6 mol/l. The solution is sprayed from a nozzle and frozen, thereby generating aluminum sulfate spheroids in the region of 1 mm in diameter. The spherical shape is maintained during drying, thermal decomposition and calcination. On completion of the freeze drying process the spheroids are amorphous but at around 300°C they

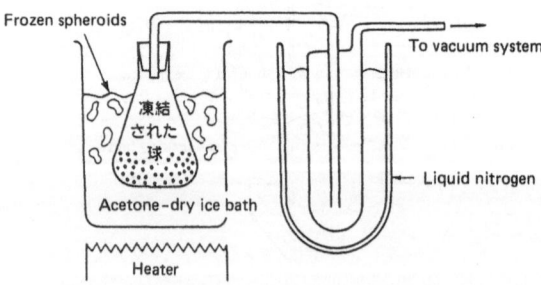

Fig. 5.22 Apparatus for drying of frozen droplets.

crystallize in the form of anhydrous aluminum sulfate. Figure 5.23 shows the thermogravimetric analysis curves in respect of the freeze dried spheroids. Gas emission analysis revealed quite clearly that up to a temperature of 500°C the gas emitted is almost all water and unless the temperature is raised to 600°C or more sulfate decomposition will not take place. Between 770°C and 860°C the alumina generated following sulfate decomposition is γ phase. At 1200°C γ phase changes to α phase. Heat treatment for 10 minutes at 1200°C is sufficient to obtain fine chained particles of several tens of nanometers in diameter and several micrometers in length.

5.7 Laser Synthesis

The process whereby a chemical reaction is induced in a liquid or a gas excited by a beam of light is known as a photochemical reaction. There already exists a substantial body of research into the effects of light excitation on, for example, atomic and molecular life spans, electronic structure and chemical properties but there is as yet virtually nothing relating specifically to the use of photochemical reaction as a means of synthesizing fine particles. Recently, however, there has been some experimentation with the use of lasers for such purposes. Figure 5.24 shows a schematic representation of apparatus for the synthesis of fine particles with a laser [26,27]. The apparatus is configured such that the direction of the laser beam and the direction of flow of the reactive gas are at right angles to each other. The laser itself is a CO_2 laser with a wavelength of 10.6 μm and a maximum output of 150 W. Defocused beam intensity is between 270 and 1020 W/cm^2 and focused intensity is

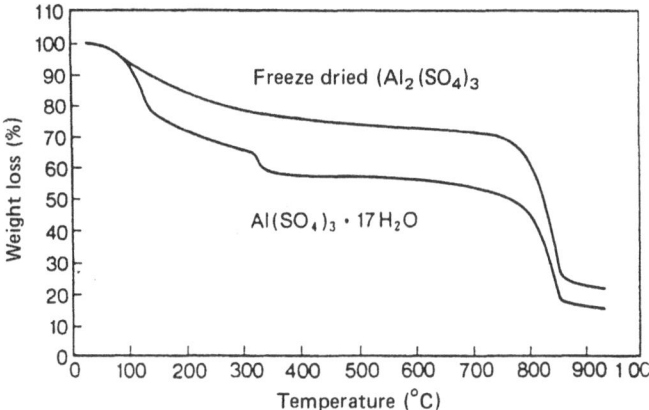

Fig. 5.23 Thermal decomposition curves of freeze dried aluminum sulfate [22].

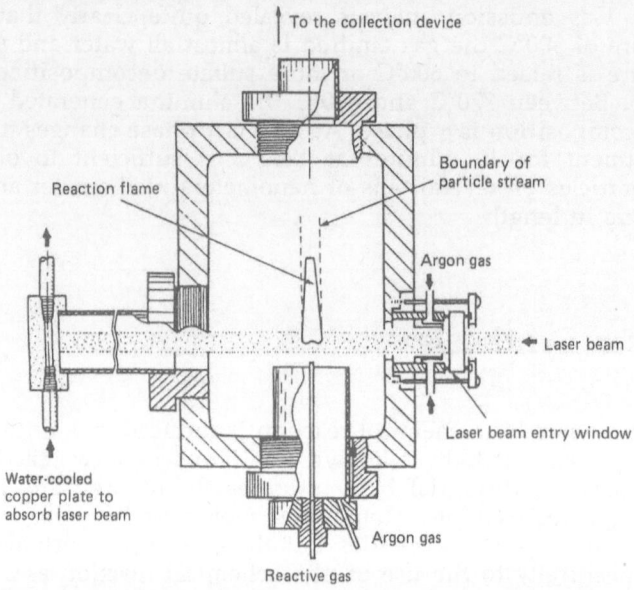

To the collection device

Boundary of
particle stream

Reaction flame

Argon gas

Laser beam

Laser beam entry window

Water-cooled
copper plate to
absorb laser beam

Argon gas

Reactive gas

Fig. 5.24 Apparatus for the laser synthesis of fine particles [26].

105 W/cm². Gas pressure in the reaction chamber is adjusted to somewhere between 0.08 and 1.0 atm and the laser beam enters the chamber through a KCl window. When the laser beam strikes the flow of reactive gas a reaction flame forms inside the reaction chamber as shown in Fig. 5.25. The fact that the flame is inclined slightly to the left indicates the presence of a stream of argon gas which is designed to prevent any movement of powder or heat in the direction of the KCl window. The particles themselves are generated inside the reaction flame, as indicated by the dotted lines in the figure, and then carried up in the argon gas stream inside the fine particle column, indicated by the dotted lines, to be finally collected on a microfilter. The use of the laser synthesis method enables the reaction space to be located at any point inside the reaction chamber, which removes the need for contact with the material of the reaction chamber itself, thereby permitting the production of ultrapure fine particles by avoiding the introduction of any impurities into the reaction. Moreover, the facility to create a reaction area with an extremely uniform, high temperature level which is effectively thermally insulated from the surrounding area makes the synthesis conditions easy to control, thereby permitting the synthesis of fine particles in monodispersion. This method has been used experimentally for the synthesis of Si, SiC and Si_3N_4 from SiH_4. The synthesis reactions used in each case were as follows:

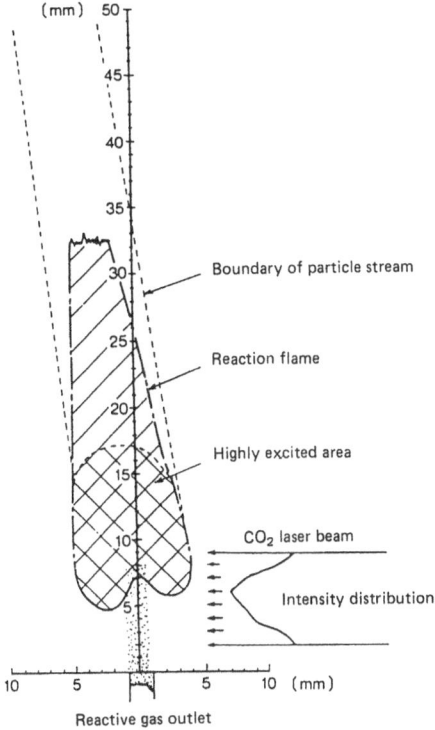

Fig. 5.25 Structure of laser generated reaction flame [27].

$$SiH_4 (g) \rightarrow Si (s) + 2H_2 (g)$$

$$3SiH_4 (g) + 4NH_3 (g) \rightarrow Si_3N_4 (s) + 12H_2 (g)$$

$$SiH_4 (g) + CH_4 (g) \rightarrow SiC(s) + 4H_2 (g)$$

$$2SiH_4 (g) + C_2H_4 (g) \rightarrow 2SiC (s) + 6H_2 (g)$$

All the fine particles obtained were spheroidal with an average diameter of approximately 50 nm in the case of Si, 10–20 nm in the case of Si_3N_4 and 18–26 nm in the case of SiC. The fine particles of Si and Si_3N_4 formed a high purity powder with a weight including oxygen of 0.1 wt% while the SiC particles were either Si-rich or C-rich. All the particles exhibited chain aggregation characteristics.

5.8 Spark Discharge

The introduction of a metal electrode into a dielectric gas or liquid and the application of an increasing voltage will lead to dielectric breakdown in the manner described by the voltage/current curve in Fig. 5.26. In other words, with the initial increase in voltage the current is also observed to increase until it reaches the corona discharge point b. When the corona discharge point has been passed the current will continue to increase naturally without any further increase in the applied voltage until it reaches the arc discharge point, which is a state of momentarily stable discharge. The transient discharge between corona and arc is known as a spark discharge. While a spark discharge is only shortlived at somewhere between 10^{-7} and 10^{-5} s the electrical potential gradient is steep between 10^5 and 10^6 V/cm, and current density rises to somewhere between 10^6 and 10^9 A/cm^2. In other words, a spark discharge releases an enormous amount of electrical energy in a very short time. This involves the generation of both a high temperature and a high level of mechanical energy at the point of discharge. This phenomenon is commonly employed in an industrial process known as electrical discharge machining, whereby the machining of a workpiece is achieved by the generation of a spark discharge between an electrode and the workpiece within a dielectric fluid such as kerosene. The machining process itself produces machine waste from both the electrode and the workpiece. The spark discharge method of fine particle production involves the positive control of this waste-producing side effect of the machining process. Experiments have been conducted into the production of alumina, for example. Figure 5.27 shows a schematic representation of the type of device used for such experiments [28]. Layers of metallic aluminum pellets are laid in a water tank into which an electrode is introduced.

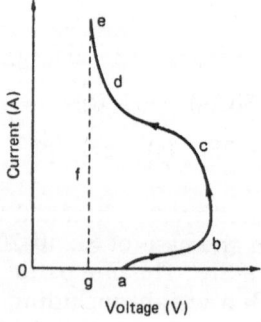

Fig. 5.26 Dielectric breakdown current/voltage curve.

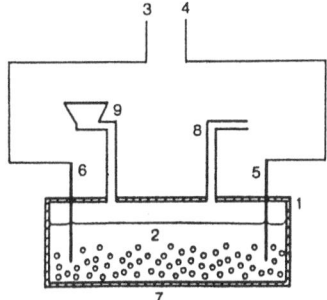

Fig. 5.27 Device for the synthesis of fine particles of alumina by the spark discharge method [28]. 1: Reaction tank, 2: pure water, 3,4: output terminals, 5,6: aluminum electrodes, 7: aluminum pellets, 8: powder and gas extraction tube, 9: aluminum pellet supply hopper.

The spark discharge taking place between the electrode and the pellets is used to produce the requisite fine particles. The reaction tank used for particle generation is 20 cm wide and 120 cm deep, the pellets flat with diameters between 10 and 15 mm, the discharge voltage is 24 kV and the discharge frequency 1200 cycles per second. During the synthesis the pellets are agitated violently by the electrical discharges, which are repeated regularly to prevent electrical fusion taking place between the pellets. The spark discharges cause exfoliation of fine surface layers of the metal which subsequently reacts electrolytically with the water. The OH radical produced by this electrolytic reaction generates a slurry of $Al(OH)_3$. This slurry undergoes a solid–liquid separation and is then dried for a period of about 24 hours to form the solids, after which fine particles of alumina powder of primary particle diameter 0.6–1 μm are obtained by comminution and calcination. The dried slurry is then comminuted and calcined to produce fine primary particles of alumina between 0.6 and 1 μm in diameter. The use of an aluminum electrode ensures a high degree of purity of the synthesized alumina.

References

1. Y. Ozaki: Ceramics, 16, 7, p. 570 (1981) (in Japanese)
2. Y. Ozaki: Ceramics, 16, 8, p. 675 (1981) (in Japanese)
3. Y. Ozaki: New Ceramic Powder Handbook, p. 79, Science Forum (1983) (in Japanese)
4. Y. Ozaki: Kagaku Kougaku, 46, 10, p. 13 (1982) (in Japanese)
5. P.K. Gallagher, J. Thomson, Jr.: J. Am. Ceram. Soc., 48, 12, p. 644 (1965)
6. D. Henings, W. Mayer: J. Solid State Chem., 26, p. 329 (1978)

7. K. Hisataka: Electronics Ceramics, 13, Summer, p. 57 (1982) (in Japanese)
8. E. Matijevic, M. Budnik, L. Meites: J. Colloid Interface Sci., 61, p. 302 (1977)
9. S. Hamada, E. Matijevic: J. Chem. Soc., 78, p. 2147 (1982)
10. Y. Ozaki: Kougyou Zairyou, 29, p. 85 (1981) (in Japanese)
11. Y. Ozaki: Nihon no Kagaku to Gijutsu, 25, 228, p. 43 (1984) (in Japanese)
12. Y. Ozaki: Electronics Ceramics, 13, Summer, p. 26 (1982) (in Japanese)
13. Nihon Funtai Kogyou Gijutsu Kyokai (ed.): Cho-biryushi Ouyou Gijutsu, p. 72, Nikkan Kogyo Shinbunsha (1986) (in Japanese)
14. Y. Ozaki: Hyoumen Kagaku, 8, p. 301 (1987) (in Japanese)
15. J.G.M. deLAU: Ceram. Bull., 49, p. 572 (1970)
16. B.J. Ingebrethsen, E. Matijevic: J. Aerosol Sci., 11, p. 271 (1980)
17. S. Kanzaki: Yogyo Kyoukaishi, 91, p. 81 (1983) (in Japanese)
18. M. Kiyama: Bull. Chem. Soc. Jpn., 47, 1646 (1974)
19. Yougyou Kyoukai Editing Committee, Kouza Subcommittee (ed.): Ceramics no Seizou Process, p. 40 (1984) (in Japanese)
20. Yoshimura, Kikugawa, Munemiya: Yougyou Kyoukaishi, 91, p. 182 (1983) (in Japanese)
21. Y. Ozaki: Shouketsuyou Ceramic Funtai no Process Gijutsu, p. 39, Yougyou Kyoukai (1980) (in Japanese)
22. D.W. Johnson, F.J. Schnettler: J. Am. Ceram. Soc., 53, p. 440 (1970)
23. Hattori, Mouri: Yougyou Kyoukaishi, 89, p. 287 (1981) (in Japanese)
24. D.W. Johnson, F.J. Gallagher, D.J. Nitti, F. Schrey: Bull. Am. Ceram. Soc., 53, p. 163 (1974)
25. A.C.C. Tseung, H.L. Bean: J. Mater. Sci., 5, p. 604 (1970)
26. W.R. Cannon et al.: J. Am. Ceram. Soc., 65, 7, p. 324 (1982)
27. W.R. Cannon et al.: J. Am. Ceram. Soc., 65, 7, p. 330 (1982)
28. W. Ishibashi et al.: Ceramics, 6, p. 461 (1971) (in Japanese)
29. F. Chou, X. Feng, J. Li, V. Lin: J. Appl. Phys. 61, pp. 3881–3882 (1987)
30. N.J. All, S.J. Milne: Br. Ceram. Trans. J., 86, pp. 113–117 (1987)

Additional References

31. S. Kratohvil, E. Matijevic: Adv. Ceram. Mater., 2, pp. 708–803 (1987)
32. F. Suzuki: Shikizai, 60, pp. 481–486 (1987) (in Japanese)
33. A. Tsugita, T. Fukushima, A. Mizuguchi, H. Nagasawa: Shikizai, 61, pp. 683–691 (1988) (in Japanese)
34. M. Yoshizumi: Silicates Industries, 7, pp. 151–157 (1984)
35. K. Akashi, T. Takeda, Y. Ozaki (ed.): "Powder preparation", Proceedings of the MRS International Meeting on Advanced Materials, Vol. 3, Materials Research Society (1989)

Chapter 6

Applications of Superfine Particles

As more is discovered about the properties of superfine particles, their potential industrial applications look increasingly promising. Applications such as sensors and magnetic tape, however, are still in their infancy. There is a good deal of potential for applications in the field of materials, especially sintered materials: in this area, materials made from superfine particles outperform the others, and their economic properties make them extremely useful. At present, however, these applications are still at the basic research stage.

The present situation regarding the application of superfine particles was to some extent predictable, from the standpoint of materials science, but these difficulties must be overcome and progress must be made. In this book, we have discussed several applications, including electronic materials, magnetic materials, optical materials, sintered materials, catalyst materials, and sensor materials.

6.1 Introduction

Probably the most consistent and also the least explored properties of superfine particles are their surface properties. Superfine particles have good potential for use as catalysts and materials for gas selection and separation, but the development of applications like these has only just begun.

The application of superfine particles still has a long way to go, and there are only two ways forward: one is to move consciously into the fields where conventional powders are used, and the other is to develop areas in which completely new substances are used. In other words, the first alternative means enhancing the material properties of superfine particles to make them superior to all other materials. To achieve this aim, low cost techniques for producing high grade superfine particles with uniform particle diameter must be developed, and techniques for stabilizing their properties must be established. The establishment of techniques involved in these applications (i.e. handling techniques) is even more vital.

The second alternative is to create new substances. To do this, the new properties arising from the superfine quality of the particles must first be studied. Changes in the state of superfine particle electrons have not yet been thoroughly researched in experimental terms, and there have been very few observations of interesting behavior in the electrons or the substances. This is a great obstacle to the applied scientists working on superfine particles. These particles will have to be comminuted still further, without contamination, and their properties studied, especially at low temperatures.

Bearing these problems in mind, we shall now look at several typical applications of superfine particles.

6.2 Use of Superfine Particles in Electronic Materials

6.2.1 Thick Film Materials

Thick film materials [1] are a good example of the electronic materials in which superfine particles are used. A mixture of glass powder and electroconductive metal powder is evenly dispersed in an organic solvent, and the result is termed a thick film paste. This film is screen printed on a ceramic plate, then fired, and used as a conductor between two circuit elements such as a resistor and a condenser, or as a contact between two electronic circuits.

The paste for film printing is made by mixing the glass powder and the electroconductive component together in a suitable ratio and dispersing the result in an organic solvent. The glass powder particles have an average diameter of between 1 and several micrometers, and are created by the mechanical pulverization method. The electroconductive particles have an average diameter of between several tens of nanometers and several micrometers, and are usually created by pyrolyzing a hydroxide.

Pastes such as this are used for passive components such as conductors, resistors, dielectrics, and insulators. Recently, advances have been made in research on pastes for active components such as oscillators and memory. Electroconductive pastes are made using finely powdered substances including Au, Pt, PdAg, Cu and Ni; substances used to make resistor pastes include RuO_2, $Bi_2Ru_2O_7$, MoO_3, LaB_6 and carbon, while those used to make dielectric pastes include $BaTiO_3$ and TiO_2. Precious metals are often used as the electroconductive component in electroconductive pastes. This is because ordinary metals oxidize, since the printed paste is fired in air. Recently, however, base metals such as Cu and Ni have come into use. This is because lowering the cost of the paste has become the chief priority in the development process, so firing is now carried out in an inert gas such as nitrogen.

The thick film pastes used in the largest quantities are electroconductors, and as we mentioned earlier, precious metals such as Au, Pt and Pd are used. Au conductors are stable and hardly react or diffuse at all even when interleaved with insulators and fired. They are therefore used to make many-layered circuits in which an electroconductor and an insulator are printed on in alternate layers. They can also be printed very finely, and up to eight semiconductor patterns can be printed in a space 1 mm wide. The special property of Au particles is that, unlike metal particles in other thick film pastes, they are almost perfectly spherical. This is probably the reason why Au paste can be printed in such fine lines. However, since precious metals are expensive, there is a large demand for low cost pastes, and to meet this demand, pastes of base metals such as Cu and Ni have been developed. Cu is particularly interesting in that it is an electroconductive material that can be substituted for Au. In future, Cu will probably be used in all the areas in which Au is used at present.

Next, let us consider the computer applications of conductive materials. Recent years have seen remarkable improvements in computer processing speed. These improvements have been achieved by a variety of means including adding extra processors, improving the device switching speed, and mounting the devices more densely, thus reducing the wiring separation between them. To increase the wiring density, circuits must be produced which have fine wiring patterns. This means that the conductor used in the wiring has a smaller cross-section, and the electrical resistance of the wiring is therefore higher. This in turn means that fewer electrical signals are propagated, and distortions occur in the waveform. If superconductive substances could be used as wiring materials in place of ordinary conductors such as Cu, this problem could be greatly alleviated.

Conventional superconductors enter the superconductive state at very low temperatures, and must therefore be cooled using liquid helium or liquid hydrogen. However, these cooling media are expensive and difficult to handle, so conventional superconductor materials present practical problems. However, high-temperature semiconductors, typified by Y–Ba–Cu–O systems, have now come to light [2], and the possibilities for producing more practical superconductors have widened considerably.

A thick film paste is made from the 90 K class superconductor Y–Ba–Cu–O system bulk superconductor shown in Fig. 6.1. The procedure is as follows: a bulk superconductor is produced using powdered yttrium oxide (Y_2O_3), powdered copper oxide (CuO) and powdered barium

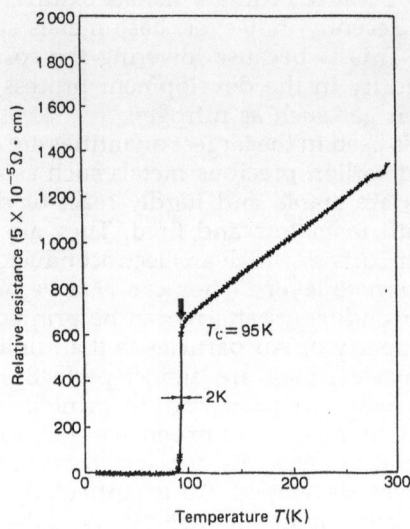

Fig. 6.1 Resistance–temperature properties of YBa$_2$CU$_3$O$_{7-\delta}$.

carbonate (BaCO$_3$) as the raw materials, and then pulverized. The pulverized powder is then evenly dispersed and mixed in an organic binder and an organic solvent, producing a paste with good elongation [3].

This paste is then printed on a high purity alumina board and heat-treated, producing a thick film circuit. The electrical resistance of the thick film material produced in this way gradually decreases as the temperature falls, reaching zero at 89 K. This value is the same as the bulk state value. To be recognized as a superconductor, a material must not only have an electrical resistance of zero; it must also be confirmed that the material shows the Meissner effect – that is, diamagnetism. This has been confirmed with samples of this thick film. The crystal structure of these samples was also studied using X-ray diffraction, and crystals as shown in Table 6.1 were obtained. The main constituent found was layer perovskite YBa$_2$Cu$_3$O$_{7-\delta}$, but patterns from Y$_2$BaCuO$_5$ and CuO,

Table 6.1 Comparison of crystal phase of bulk and thick film (X-ray relative intensity)

Sample	YBa$_2$Cu$_3$O$_{7-\delta}$	Y$_2$BaCuO$_5$	CuO
Bulk	60	19	21
Thick film	47	13	40

which show high resistance, were observed. The thick film was found to contain a higher proportion of CuO than the bulk state.

Superconductor thick film circuits can be used in electronic appliances such as IC boards of various types, and in hybrid circuits. In these cases, high speed wave signals can be propagated with minimal loss, an improvement on normal conductor wiring. Also, because the patterns can be made finer, densely-wired high speed circuit boards can be achieved at the temperature of liquid nitrogen, as we said before. Another major advantage of superconductor circuits is that power lines can be made finer, taking advantage of the low electricity loss and low heat generation: this technology opens up many possibilities for the future.

6.2.2 Mounting Materials

Semiconductor integration increases every year, moving from LSI towards VLSI and even ULSI. The new higher integration densities have made it impossible to use flat circuits for all the interconnections between ICs or between an IC and an external peripheral circuit. To raise mounting density in hybrid ICs, it has become necessary to use three-dimensional wiring instead of the two-dimensional wiring used for flat surfaces [4].

This can be done by various methods, but the one which is attracting most attention at present is the green sheet lamination method. In this method, several alumina green sheets printed with earth, input/output and signal circuits are piled on top of one another. They are linked together by through-holes between the circuits on each sheet. The number of layers varies according to the circuit system, ranging from a few to over thirty. When the layers have been sintered in a reducing atmosphere at a temperature well in excess of 1000°C, the LSI is mounted on the topmost layer. The layer structure is as shown in Fig. 6.2. Recently, ceramic sheets have been used as insulators too, and structures have been produced which included condensers, resistors and other components built into the board. This laminated system is sintered at a high temperature, so the position of the through-holes, the external dimensions, and other factors are altered by shrinkage. Also, the green sheet needs to be extremely thin. One solution to these problems is to use superfine particles of high purity alumina with particles of a uniform size.

Using superfine alumina particles formed by the aluminum alkoxide method, high performance super-thin alumina boards, between 30 and 100 μm thick, have been developed [5]. The alumina film is produced using the sol–gel method, as shown in Fig. 6.3. This process is as follows: an evaporation operation is first carried out, using metal and alcohol as the starting materials, and a high purity aluminum alkoxide is synthesized. The alkoxide is then evenly mixed in the liquid state, and the desired ceramic composite is formed. Next, water is added to the alkoxide mixture solution, hydrolysis is carried out, superfine particles are formed, and a bermite sol is formed by deflocculation. An organic binder is added to

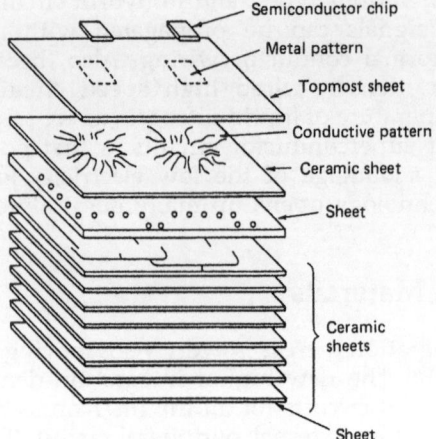

Fig. 6.2 Ceramic laminated structure.

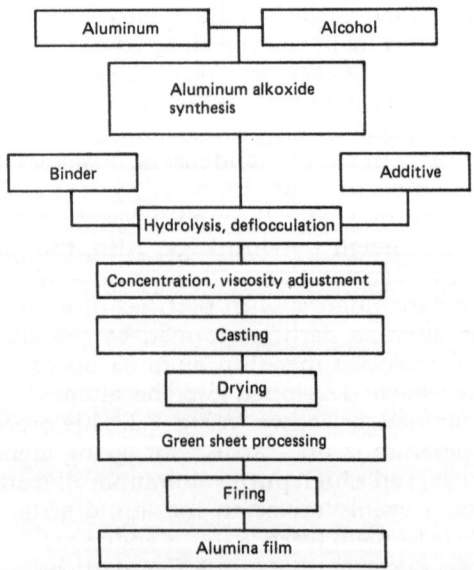

Fig. 6.3 Alumina film production process.

this sol, the result is concentrated, and the viscosity adjusted, after which the casting and drying processes are carried out, and a green sheet is obtained. The widely used doctor blade method can be used in the casting process, but the fluidity of the sol must be borne in mind when rigging up the casting head.

Using this new process, it is possible to produce homogenous, fine alumina film between 30 and 100 μm thick, without going from the raw material stage to the powder state. Moreover, a fine alumina film can be obtained with a sintering temperature between 100 and 150°C lower than usual [6].

The properties of the alumina film are shown in Table 6.2. In the microstrip lines used in high-frequency circuits, when the permittivity of the board is 10, and the impedance of the power line is 50 Ω, the thickness of the line must be almost the same as the thickness of the board. Therefore, by using alumina film, the strip lines can be made finer, and LSI chips for high-frequency use can be mounted at high densities. Also, the high Young's modulus of alumina and the thinness of the alumina film mean that it can be used for applications such as diaphragms. Alumina film can also be used in a wide range of applications including sensor boards and, when combined with metals, radiating boards.

Table 6.2 Characteristics of alumina film

a Electrical properties

Purity (%)	Permittivity (1 MHz)	tan δ (1 MHz)	Specific volume resistivity (Ω·cm)	Dielectric strength (kV/mm)
96	9.5	5×10^{-4}	$> 10^{15}$	> 30
97	9.7	4×10^{-4}	$> 10^{15}$	> 30

High-frequency characteristics (1 GHz)
  Permittivity: 9.31
  tan δ: 3.7×10^{-4}

b Mechanical strength characteristics

Thickness (μm)	Bending strength (kgf/cm^2)	Young's modulus (kgf/cm^2)	Breaking load[a] (kg)
60	6000	3.2×10^6	0.07
80	5900	3.3×10^6	0.11
100	5500	3.4×10^6	0.13

[a]Measurement conditions:
  Measurement method: three-point bending
  Span: 20 mm
  Measured sample: 30 × 10 mm

The structure of a ceramic multi-layer wiring board using alumina film is shown in Fig. 6.4. The conductors and resistors on this board were screen-printed onto the alumina film in which through-holes have been made. A thick film paste of a substance such as Ag, Ag–Pd, Ag–Pt, and Au – now generally used in hybrid ICs – can be used as the conductor material. If a thick film conductor bump is used for the path between the boards, non-crystalline glass can also be used to link the boards mechanically.

Many-layered boards using alumina film have the following advantages over boards produced by the conventional green sheet lamination method.

First, the conventional green sheet lamination method requires the use of tungsten or molybdenum, which have a lower conductivity than ceramic green sheets. Also, the green sheets have to be fired in a reducing atmosphere such as hydrogen, in order to prevent the conductor oxidizing. Recently, low temperature firing boards have been developed which can be fired simultaneously with thick film pastes such as Au and Ag–Pd in oxidizing atmospheres at a temperature between 800 and 1000°C, but there are various problems: because the boards shrink when the green sheets are fired, it is difficult to control the dimensions, so the pattern design has to allow for shrinkage; besides, it is not clear that the glass component, introduced to lower the sintering temperature, is compatible with conventional thick film pastes.

Many-layered wiring boards using alumina film, however, have advantages: designing the patterns is easy, because the boards do not shrink, and the thick film materials used in conventional ceramic HICs can be used without modification. Furthermore, because alumina film is used, the individual layers can be kept at the same thickness as those used in conventional green sheet laminated boards.

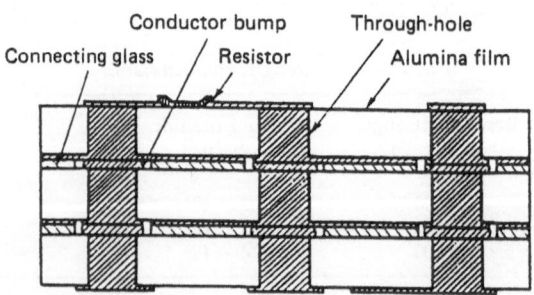

Fig. 6.4 The structure of a ceramic multi-layered wiring board.

6.3 Use of Superfine Particles in Magnetic Materials

6.3.1 Magnetic Recording Media

Now that more information is being recorded, and recording devices are becoming progressively miniaturized, the demand for higher recording density is more acute than ever before. As is well known, the vertical magnetization recording method has recently been attracting a good deal of attention as a way of meeting this need. This method affords a much greater recording density than conventional methods of recording parallel to the surface of the recording medium. A thin film of a metal alloy such as Co–Cr is used as the vertical magnetic recording medium, and is applied using a vacuum process such as sputtering or vapor deposition. R. & D. in this field is now progressing rapidly. However, the majority of magnetic recording media, such as audio and video, now in practical use, are coated magnetic recording media with recording parallel to the surface, wherein acicular magnetic powder – typically Co adhered γ-Fe$_2$O$_3$ – is dispersed in a binder and applied to a base film. If magnetic recording media capable of being vertically magnetized could be produced by the application method, the past store of technical expertise regarding coated magnetic recording media could be widely utilized, and this would be extremely convenient. However, conventional magnetic powders, such as γ-Fe$_2$O$_3$, CrO$_2$, Co adhered γ-Fe$_2$O$_3$, and metal (Fe), all consist of acicular particles, and magnetization only remains in the longitudinal direction. In order to produce vertical magnetized recording media using these acicular particles, the particles must be positioned perpendicular to the base surface, which is likely to be extremely difficult.

The powder which has been newly developed to replace the acicular magnetic particles is barium (Ba) ferrite powder. Fine particles of barium ferrite are hexagonal plate crystals, as shown in Fig. 6.5, and one of their characteristics is that they are easily magnetized in the direction perpendicular to the wide surface. Therefore, when these hexagonal plate crystals are aligned, by application, in such a way that their hexagonal surface is parallel with the base surface, they constitute a vertically magnetic recording medium.

Broadly speaking, there are two methods of producing fine crystals of Ba ferrite: the glass crystallization method [7] and the hydrothermal synthesis method [8]. Here, we shall describe the Ba ferrite particles produced by the glass crystallization method, and the recently developed acicular Ba ferrite particles.

The process of producing Ba ferrite particles by the glass crystallization method is as shown in Fig. 6.6. In this process, Ba ferrite is mixed with a substance used to make glass, typically B$_2$O$_3$, and dissolved at a temperature of 1300°C or over. The result is rapidly cooled and (non-

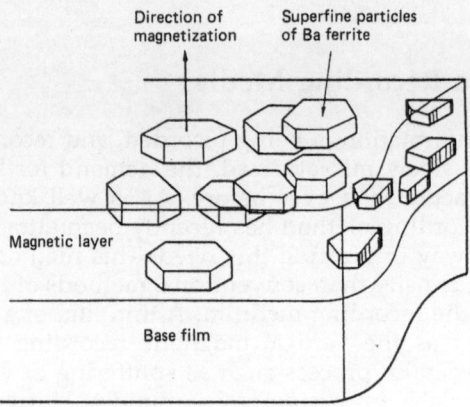

Fig. 6.5 Ba ferrite coated perpendicular magnetic recording medium.

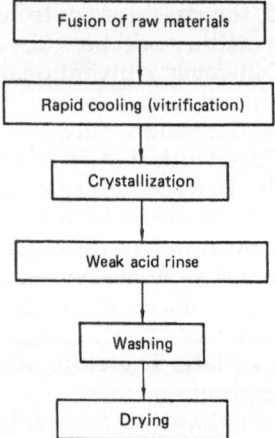

Fig. 6.6 Production process for superfine Ba ferrite particles using the glass crystallization method.

crystalline) glass. This glass is reheated at a temperature of 700°C or over, and the Ba ferrite is crystallized. By dissolving the excess glass using acid or salt water, and then rinsing and drying the result, Ba ferrite particles can be obtained. The characteristics of this method are that it gives an even granularity distribution, because when the crystals are formed, the Ba ferrite particles are deposited from the glass phase, where the viscosity is high, and that Ba ferrite powder with good dispersion is obtained because the layer of glass between the Ba ferrite particles stops the particles flocculating. Moreover, Fe displacement can be carried out easily for coercive force control. Superfine magnetic particles with a diameter of 0.1 μm and under can be obtained. This is a truly excellent method for producing magnetic powder for vertical magnetic recording media. The typical characteristics of Ba ferrite powder obtained by the glass crystallization method are shown in Table 6.3.

As mentioned above, superfine particles of Ba ferrite have many excellent characteristics as vertical magnetic recording media, but when they are dispersed in the coating material and applied to the base, it is difficult to apply them in such a way that the magnetic axis is always perpendicular to the base, as shown in Fig. 6.5. That is to say, superfine lamellar particles are more difficult than superfine acicular particles to orient and apply in such a way that the direction of the applied magnetic field is the same for each particle. Experiments have also been carried out on the production of Ba ferrite particles with the same acicular form as conventional γ-Fe_2O_3 particles [9].

The acicular Ba ferrite particles are produced using the topotaxy reaction between acicular α-FeOOH particles and $BaCO_3$ particles. In other words, as shown in Fig. 6.7, the crystal axis of the surface of the acicular α-FeOOH is [100], and the Ba^{2+} ions of the $BaCO_3$ diffuse from their surface; Ba ferrite is then formed inside the acicular FeOOH, whereupon the crystal axis of the surface of the acicular particles becomes [0001], the value for hexagonal Ba ferrite crystals. Then, inside the acicular particles, as shown in Fig. 6.8, many fine crystals of Ba ferrite are formed, the magnetic axis of each crystal being perpendicular to the apse line. When these acicular Ba ferrite particles are applied to a plastic base, vertical magnetic recording becomes possible.

The actual production procedure is as shown in Fig. 6.9. First, acicular α-FeOOH particles are dispersed in a $BaCl_2$ solution, $NaHCO_3$ is added, and colloidal $BaCO_3$ particles are deposited on the surface of the α-FeOOH particles. The NaCl contained in the colloidal $BaCO_3$ is washed in salt water, filtered, and then dried, a small quantity of B_2O_3, P_2O_5, or Bi_2O_3 – oxides with low melting point – is added. The result is fired at between 780 and 840°C, and then slowly cooled. This produces acicular Ba ferrite. What is most important in this process is that the acicular

Table 6.3 Typical characteristics of Ba ferrite superfine particles

Average diameter	53 nm
Average depth	18 nm
Specific surface area	31 m²/g
Coercive force	400–1500 Oe
Saturation magnetization	> 55 emu/g
Curie temperature	320°C

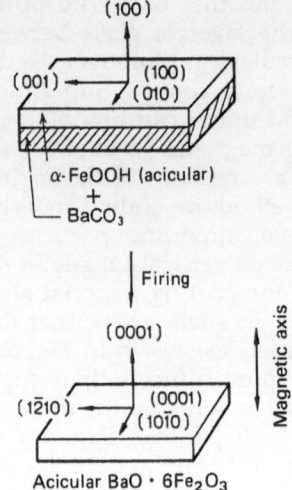

Fig. 6.7 Formation of acicular Ba ferrite using the topotaxy reaction.

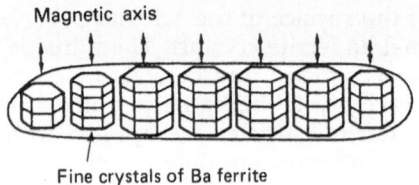

Fig. 6.8 Internal structure of acicular Ba ferrite particles and direction of magnetic axis.

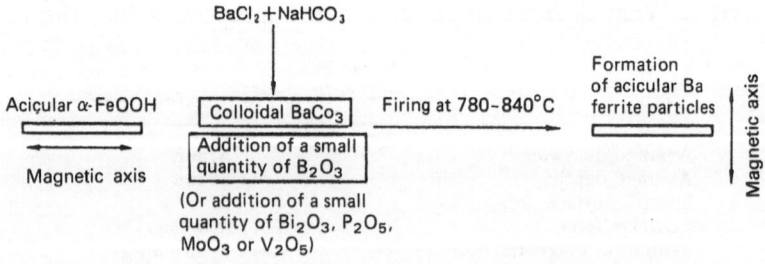

Fig. 6.9 Acicular Ba ferrite particle production process.

particles should not stick together during firing. When the firing temperature reaches 850°C or higher, adhesion occurs between the acicular particles. The Ba ferrite obtained in this way has the following magnetic characteristics: saturation magnetization: 32 emu/g; coercive force: 2000 Oe; Curie temperature: 443°C. The saturation magnetization of these acicular Ba ferrite particles is about half that of conventional Ba ferrite particles, and the raising of the saturation magnetization value remains a problem to be tackled in the future. The main features of the Ba ferrite magnetic recording media obtained in this way are as follows [10].

1. By replacing the Fe with Co–Ti, it is possible to produce a medium which allows high density recording with comparatively low coercive force. Therefore, ferrite ring heads – which have excellent resistance to wear – can be used, and metal heads are unnecessary.

2. Since the media can be given the desired level of coercive force by means of the replacement mentioned above, this level can be adjusted to suit new applications. This makes for good compatibility with other magnetic recording media.

3. The head contact, direction, and durability of these media can be ensured using the same techniques used for conventional coated media.

4. When these media are used in circular flexible disks such as floppy disks, as shown in Fig. 6.10, there is no rotation cycle modulation.

5. These media are not subject to chemical changes, and remain stable even after repeated playbacks.

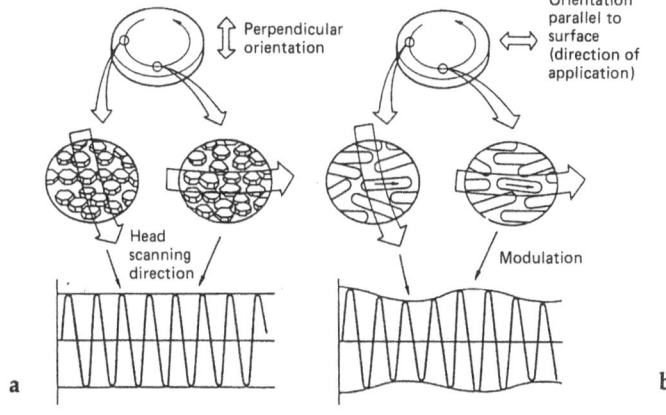

Fig. 6.10 Playback output envelope in circular flexible disk. **a** Ba ferrite disk, and **b** conventional disk.

6. These media can be coated at high speeds, and lend themselves extremely well to mass-production.

Because they have so many advantages, it is likely that Ba ferrite magnetic recording media will eventually be used in a wide range of applications in the area of high density magnetic recording. In video recording, they may well be used in applications such as 8 mm video tape and high quality video tape, and in the field of audio recording, in applications such as digital audio tape (DAT). Also, when used in 3.5 inch floppy disks, these media will make it possible to produce disks with a recording capacity of 4 M on each side – four times the capacity of the 3.5 inch 1 M disks used at present [11]. Another promising applications area in the field of floppy disks is electronic stills camera media.

6.3.2 Magnetic Fluids

A magnetic fluid [12] is produced by stably dispersing strongly magnetized superfine particles in a liquid such as water or oil. A magnetic fluid is not subject to flocculation and settling in normal gravity or in magnetic fields, and behaves to all appearances as a strongly magnetic body. A magnetic fluid, therefore, can be thought of as a fluid which can be controlled magnetically. The structure of a magnetic fluid is as shown in Fig. 6.11, and it is produced as follows: magnetite (Fe_3O_4) particles with a diameter of about 10 nm are covered with a surface active agent such as oleic acid, and made to disperse in the liquid

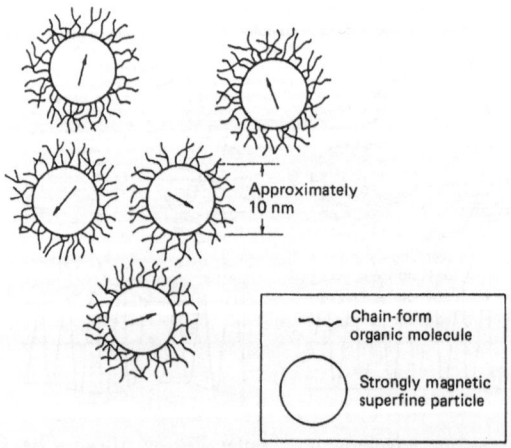

Fig. 6.11 Microscopic structure of a magnetic fluid.

Table 6.4 Typical methods for producing magnetic fluids

Category	Methods	Remarks
Pulverization method	Magnetite is pulverized over a long period of time in an organic phase containing oleic acid	1965. Papell Long period of time needed for production
Adsorption in aqueous solution – organic phase dispersion method	Oleic acid ions are adsorbed in wet magnetite in aqueous solution, washed, dehydrated, and subjected to dispersion processing	1966. Shimo, Iisaka et al. Allows large-volume production at low cost
Deflocculation method	Alkyl is added to a liquid containing Fe^{2+} and Fe^{3+}, then added to heated kerosene containing oleic acid	1972. Khalafalla

phase [13]. These particles are supported by the heat movement of the molecules of the medium, and form a stable colloid without flocculating or settling. The chain-form organic molecules used as the surface active agent have, for example, carboxyl radicals (–COOH), phosphone radicals ($-PO_3H_3$), sulfonic acid radicals ($-SO_3H$) or amine radicals ($-NH_2$) as their terminal group, and the tail is made up of hydrocarbon chains with mediophilic characteristics, between 2 and 4 nm long. These give mediophilic characteristics to the superfine particles of Fe_3O_4.

Table 6.4 shows methods reported to date for producing magnetic fluid using nonpolar solvents as the dispersion medium. Table 6.5 shows the types of solvent used to make magnetic fluid, and their features. The hydrocarbons which can be used are saturated hydrocarbons such as hexanes, heptanes and octanes, or kerosene. The esters and diesters which can be used include tetra nonyl silicic acid and 2-ethyl hexyl adipic acid.

Another important factor from the viewpoint of applications is the

Table 6.5 Properties of solvents

Solvent	Properties
Water	Low cost, low viscosity
Hydrocarbons	Low cost, low viscosity
Esters	Cold resistant
Diesters	Low vapor pressure, low viscosity
Polyphenyl ether	Low vapor pressure, low viscosity
Saturated hydrocarbons	Poor combustibility, poor solubility

type of magnetic colloidal dispersoid used. Almost all the magnetic fluids now commercially available use Fe_3O_4 as the dispersoid, although other spinel-types such as Mn, Ni, Mn–Zn and Ni–Zn by themselves or as complex ferrites, can also be used comparatively easily. Co, Fe and Ni particles are among the metallic colloids which can be dispersed. When Co particles are used, dicobalt octacarbonyl is pyrolyzed in a polymer such as acrylonitrile, and then dispersed in a hydrocarbon solvent while ultrasound is simultaneously applied; a magnetic fluid with Co particles with a diameter of about 5 nm is then obtained [14]. In another method, superfine particles of Fe alloy are dispersed in a hydrocarbon solvent using spark erosion [15]. In yet another, Ni particles are generated by degrading Ni tetracarbonyl using ultraviolet rays, or by the oxygen reduction of a nickel compound, and dispersed in a hydrocarbon solvent containing a surface active agent.

Magnetic fluids have the following magnetic characteristics: because the magnetic particles are below the critical size, they show superparamagnetism, as is well known, and the magnetization curve does not show hysteresis. The particle diameter at which a strongly magnetic body becomes a superparamagnetic body varies according to its properties, but for Fe_3O_4, the critical diameter is said to be about 30 nm. The magnetization of a Fe_3O_4 colloid with a particle size of about 10 nm in a magnetic fluid is presumed to undergo thermal fluctuation in a solid, and this is in accordance with the fact that its magnetization appears to be extremely small. Therefore, as shown in Fig. 6.12, the phenomenon of magnetic hysteresis does not occur.

Table 6.6 shows a range of magnetic fluids, and these are used in various applications [16]. A typical application is in rotating axis seals. As shown in Fig. 6.13, the basic structure of the rotating axis seal is as follows: a magnetic circuit made up of the annular pole-piece and permanent magnet is positioned around the rotating axis, and a magnetic fluid is held in position by the magnetic field generated in the space between the rotating axis and the pole-piece.

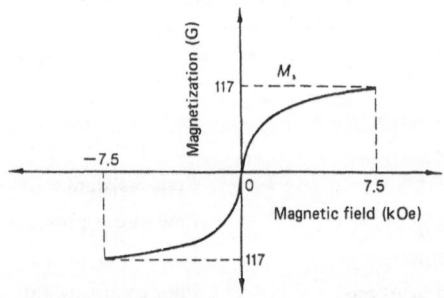

Fig. 6.12 Magnetization curve for a magnetic fluid.

Table 6.6 Properties of a range of magnetic fluids (ferricolloids made by Tohoku Kinzoku Kogyo)

	W-35	HC-50	DEA-40	DES-40	NS-35 A	PX-10
Appearance	Black liquid	Blackish-brown liquid	Black liquid	Black liquid	Black liquid	Black liquid
Magnetization (8 kOe) G	360	420	400	400	350	100
Specific gravity (25°C)	1.35	1.30	1.35	1.35	1.28	1.05
Viscosity (25°C) cps	30	30	400	600	900	1000
Boiling point (760 mmHg)°C	100	180–212	335	377	251	260
Pour point (°C)	0	−27.5	−72.5	−62	−10	−13
Ignition point (°C)		65	192	215	225	240
Vapor pressure (mmHg)			2.5(200°C)	0.5(200°C)	$7 \times 10^{-10}(20°C)$ $5 \times 10^{-3}(150°C)$	
Dispersion medium	Water	Kerosene	Diester	Diester	Alkyl naphthalene	Mineral oil
Typical applications	Specific gravity sorters	Specific gravity sorters	Seals Dampers Bearings	Seals Dampers Bearings	Vacuum seals Dampers Bearings	Dampers

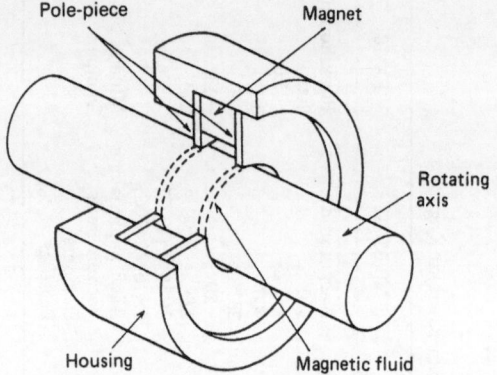

Fig. 6.13 Structure of magnetic fluid seal.

A magnetic fluid seal is completely tight and generates no friction and almost no heat. The structure is also simple. Furthermore, the pressure resistance can be increased by multistaging, and the seal can be used in a vacuum and under high pressure. In a conventional vacuum vapor deposition device, it is impossible to obtain a complete vacuum, due to factors such as the oil spilt when the rotating axis is introduced, and the dust generated by friction. With the magnetic fluid seal, however, these problems have been solved.

Electricity storage flywheels are also being explored as a means of using energy effectively: using a magnetic fluid seal, the interior of the flywheel housing can be depressurized to about 10^{-1} Torr, the friction resistance of the bearings and flywheel can be minimized, and the dynamic loss due to rotation can be reduced.

In fixed magnetic disk devices, as the memory density increases, the clearance between the head and the disk surface decreases to between 0.2 and 0.5 μm, thus preventing lubricating oil which has evaporated from the bearings, and, of course, dust from outside, from settling on the surface of the disk. Because of this dust-excluding property, magnetic fluid seals like the one shown in Fig. 6.14 are used as dust seals.

Besides seals, another typical example of the application of magnetic fluids is as follows: a magnetic field gradient is applied to a magnetic fluid, and when a non-magnetic body is placed in the fluid, buoyancy is generated in the opposite direction to the magnetic field. Some specific gravity sorting devices use this force to separate out metals such as Al, Cu and Zn contained in automobile scrap, and some dampers make use of the fact that a magnetic field applied to a magnetic fluid causes an increase in the apparent density of the fluid. The structure of a speaker damper is shown in Fig. 6.15. Furthermore, a magnetic fluid will form spike-shaped protrusions in the direction of the lines of magnetic force, and this property is utilized in devices such as ink jet printers, where ink is the magnetic fluid.

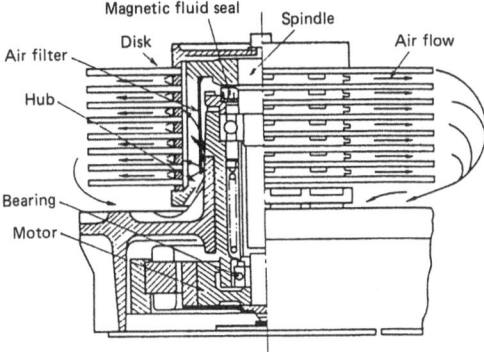

Fig. 6.14 Magnetic fluid used in dust seal in magnetic disk device.

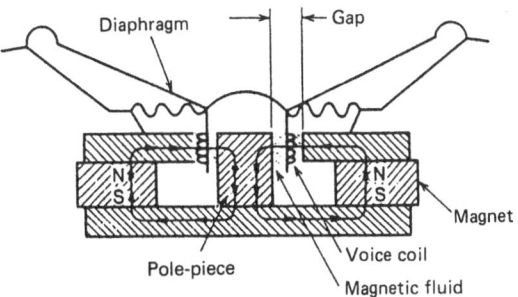

Fig. 6.15 Structure of damper used in speaker.

Magnetic fluids, with their special properties, are used in many other applications besides the examples given here.

6.4 Use of Superfine Particles in Optical Materials

6.4.1 Optical Fiber Materials

Quartz glass is the optical material which has attracted the most attention as an optical fiber material. A great deal of research is now being carried out into methods of synthesizing high purity quartz glass, especially methods involving superfine particles.

Some methods for producing superfine glass particles use gas phase reactions, and some use liquid phase reactions. In gas phase reactions, fine glass particles, between several tens and several hundred nanometers in size, are obtained by flame hydrolysis of a silicon chloride. Since this method yields very low-loss optical fiber glass, it is now in practical use for producing the basic material for low-damage optical fibers used in communications applications.

In liquid phase reactions, silicon alkoxide is generally used as the raw material. When silicon alkoxide is hydrolyzed, superfine particles with a size of up to several tens of nanometers are generated. Usually, these superfine particles are in a sol state, dispersed in a liquid phase, but the sol liquid loses its fluidity as the reaction progresses, and becomes a gel. By drying and sintering this gel, transparent glass is obtained. The sol–gel method is very promising in terms of cost, since it allows sintering to be carried out at relatively low temperatures. We shall now look at some techniques for producing glass – to be used in optical fiber materials – by the sol–gel method [17].

Broadly speaking, there are two types of methods which use a solution as the raw material to synthesize glass by the sol–gel process. In the first category, the raw material is inorganic silicate solution, and in the second category, it is an alcohol solution of metal alkoxide.

In the first type of method, an inorganic silicate solution of glass or potassium silicate is dissolved in an alkaline solution such as an ammonia solution, and then made into a weakly acidic silicic acid gel. The regular-shaped molding of gel is gradually pyrolyzed and heat-hardened to produce glass.

The second type of method is known as the alkoxide method: a non-crystalline inorganic oxide is produced using a metal alkoxide as the raw material. In this method, a metal alkoxide solution is hydrolyzed at room temperature or thereabouts, and a sol is formed by the polycondensation reaction which occurs at the same time. The reaction is taken further, and the sol is turned into a gel, which is then pyrolyzed at a low temperature to produce glass. The pyrolysis can be carried out at a much lower temperature than the melting point of the oxide. This method could be termed a kind of sol–gel method, and since glass is obtained by pyrolysis at a low temperature, it could also be called a low temperature glass synthesis method.

The practical details of this method are as follows [18]. A metal alkoxide such as ethyl silicate $Si(C_2H_5O)_4$ is mixed with water and ethyl alcohol, then this mixture is hydrolyzed at an even temperature, yielding a transparent agar-like gel. The regularly-shaped gel is heated, driving out the alcohol and water, and silica glass with a three-dimensional SiO_2 structure is obtained. This synthesis method is shown in Fig. 6.16. The chemical formula for this reaction can be written as follows:

$$Si(C_2H_5O)_4 + 4H_2O \rightarrow Si(OH)_4 + 4C_2H_5OH \uparrow$$

$$Si(OH)_4 \rightarrow SiO_2 \text{ (silica glass)} + 2H_2O \uparrow$$

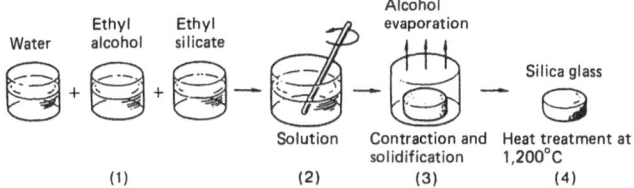

Fig. 6.16 Synthesis of silica glass by the sol–gel method.

The above formula describes an absolutely pure reaction; in practice, the reaction is more complicated. The features of this method and the possibilities it opens up are listed below [19].

1. High purity: since the solution which forms the raw material can be purified to a high degree, a highly pure glass can be obtained. This represents a great advantage over high temperature fusion methods, wherein, if a container is used, the substance of the container contaminates the substance being fused.

2. High homogeneity: since a low-viscosity material (a liquid) is used, the glass obtained is highly homogenous. Also, when a glass containing two or more components is being produced, the alkoxide is mixed in a liquid state, or as a solution, so the materials mix together evenly at the molecular level, and the result therefore has good homogeneity.

3. New compositions: there are some glass compositions which cannot be used with the fusion method, since liquid phase immiscibility (the phenomenon whereby two liquids with different compositions separate) or crystallization during cooling make it impossible to obtain a homogenous product. The alkoxide method, however, is not subject to these problems. Therefore, glass can be made with new compositions which could not be obtained before.

4. Energy saving: using this method, glass is synthesized at relatively low temperatures of up to 1200°C, and this saves energy. When silica is formed by the high temperature fusion method, using natural rock crystal as the raw material, a high temperature of close to 2000°C is required.

5. Combination with other substances: since this method does not require heating to high temperatures, the glass can be combined with metals, plastics and other materials which cannot withstand very high temperatures.

So far, we have only dealt with the advantages of the alkoxide method, but its disadvantages are as follows: the raw materials are not cheap; a substantial proportion of the water content remains in the product; the gel undergoes a large reduction in volume during shrinkage and solidification, and is therefore prone to cracking; generally speaking, it takes a long time to obtain the glass.

The sol–gel method described above is extremely attractive as a glass production process, but in order to obtain glass for low-loss optical fibers, various problems in the gel sintering process must be tackled: a large quantity of organic matter such as water, hydroxide radicals or alcohol, is adsorbed at the surface of the gel particles; also, the specific surface area of the gel is very large – between 500 and 800×10^3 m^2/kg, so it has to be by given heat treatment to remove pores. The loss characteristics of optical fibers which have undergone these treatments, and which have been produced under optimum conditions, are as shown in Fig. 6.17: the characteristics of the treated samples are markedly better than untreated samples, and optical fibers which lose 10 dB/km or less at wavelengths of 0.6 μm and over can be obtained.

6.4.2 Infrared Reflection Film Materials

Heat ray reflection films allow light rays in the visible radiation band to pass through, while reflecting infrared rays. The energy-saving potential of these films is now attracting considerable interest, and the films are being applied in many areas. Sixty-nine per cent of radiation from an incandescent light bulb is infrared rays, and many attempts have been made to reuse this energy effectively. However, this idea could not be put into practice until the advent of low cost practical materials which allow visible light to pass through while effectively reflecting infrared rays. Recently, TiO_2–SiO_2 laminated interference films produced by the alkoxide method have been used as infrared reflecting films: we shall now describe a typical example [20].

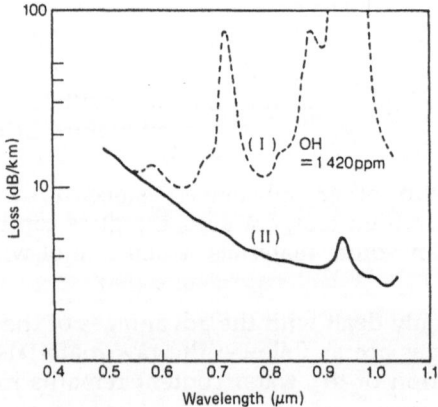

Fig. 6.17 Transmission loss wavelength characteristics of optical fibers produced by the sol–gel method. *I*, untreated; *II*, treated.

Generally speaking, when a substrate is coated with an infrared reflection film, this is done for one of several reasons. If the substrate is made of glass, the coating may serve to increase its chemical durability (resistance to water and acid) and the caustic resistance, to preserve its mechanical strength, adjust its reflectivity, add color, increase its electroconductivity, and so on, depending on the kind of glass used. If the substrate is made of metal, the coating may serve to increase the metal's rust resistance, or acid resistance, or to give it insulation. For a plastic substrate, the film may serve to protect the surface or adjust the reflectivity.

The advantage of producing the coating by the alkoxide method is that dipping can be used. The substrate can be dipped in the solution at normal temperature, and heating can then be completed at comparatively low temperatures: the whole process can therefore be carried out at temperatures lower than the point at which a glass substrate softens, or the point at which a metal substrate oxidizes or softens. There may even be cases where the substrate is ready to use after merely being dipped in alkoxide solution at normal temperature and then dried so that the adhering solution becomes a gel.

As shown in Fig. 6.18, infrared reflection films include metal–dielectric composite films, dielectric films and laminated interference films. Table 6.7 shows the composition, materials and production methods for the principal infrared reflection films. As shown in Table 6.8, the characteristics and problems of each type of film are as follows: metal–dielectric composite films possess really excellent optical properties, but their heat resistance is poor – they can only withstand temperatures up to 200°C. Conversely, dielectric films are heat-resistant, but their optical properties are not very good. On the other hand, laminated interference films have excellent resistance to heat, and their optical properties can also be enhanced, by using titania (TiO_2) or silica (SiO_2) to form the film.

A good way to make metal oxide crystals with a high melting point, such as TiO_2 and SiO_2, is the superfine particle oxide adjustment method which involves hydrolyzing a metal alkoxide solution. The formation of films by the dipping method using metal alkoxide solution is carried out by the basic process shown in Fig. 6.19.

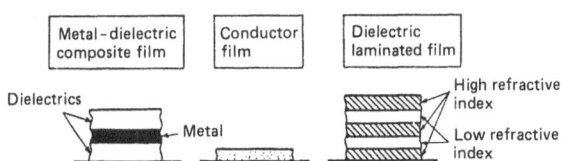

Fig. 6.18 Structure of infrared reflecting film.

Table 6.7 Composition, materials, and production methods for the principal infrared reflecting films

Form	Composition	Materials	Production method
Thin metal film	Au, Ag, Cu	Metal	Vacuum vapor deposition
Transparent electroconductive film	SnO_2, In_2O_3	Metal Oxide Other compound	Vacuum vapor deposition Sputtering Spraying
Laminated interference film (1) (Dielectric–dielectric)	$ZnS–MgF_2$ $TiO_2–SiO_2$ $Ta_2O_5–SiO_2$	Organic metal compounds Oxides Other compounds	Vacuum vapor deposition CVD Dipping
Laminated interference film (2) (Dielectric–metal–dielectric)	$TiO_2–Ag–TiO_2$ $TiO_2–MgF_2–Ge–MgF_2$	Oxides Metals	Vacuum vapor deposition Sputtering

Table 6.8 Features of infrared reflecting film

	Metal–dielectric composite film	Conductor film	Dielectric laminated film
Optical properties	◎	△	○
Heat resistance	×	○	◎
Cost	△	◎	×

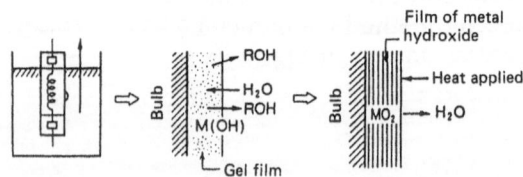

Fig. 6.19 Formation of film from metal alkoxide solution.

then removed at a fixed speed, so that a thin film is formed on the surface of the glass bulb.

2. The alcohol used to make the solution is evaporated out of the film. The alkyl radicals which have reacted with the moisture in the air and

The alkyl radicals which have reacted with the moisture in the air and combined with the metal evaporate as alcohol, and a clear film of gel is formed. The reaction is as follows

$$M(OR)_n + nH_2O \rightarrow M(OH)_n + nROH$$

Here, M is the metal, and R is the alkyl base.

3. When this gel film is fired in air, the following reaction takes place, and a metal oxide film is formed

$$M(OH)_n \rightarrow MO_{n/2} + \frac{n}{2} H_2O$$

The thickness of the films formed by these processes depends on the viscosity of the solution, the concentration of the metal oxide, and the speed of withdrawal from the solution. However, as shown in Fig. 6.20, if the concentration and viscosity of the solution are kept at a fixed level, the thickness of the film can be controlled to within several nanometers comparatively easily by controlling the withdrawal speed.

In interference films obtained as described above, the refractive index of the film is an important factor. The refractive index of such films is strongly influenced by the firing conditions. Figure 6.21 shows the relationship between the firing temperature and the refractive index for TiO_2. If the firing temperature is 600°C or over, the refractive index will be 2.1 or over, and the result can be used as an interference film. The

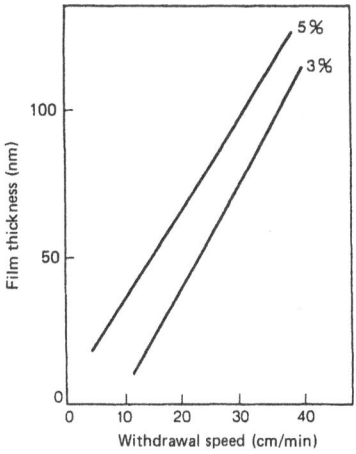

Fig. 6.20 Relationship between TiO_2 film and withdrawal speed.

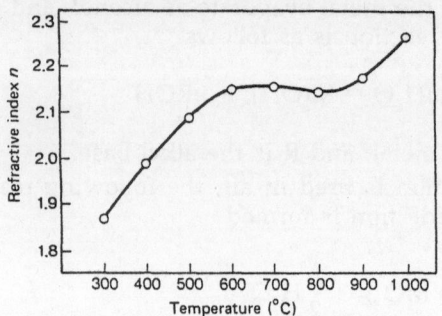

Fig. 6.21 Relationship between firing temperature and refractive index for TiO_2 film.

increase in the refractive index at 900°C and over is probably due to the fact that the film, originally a non-crystalline substance, has partially crystallized.

As the crystallization progresses, the transmissivity to visible light decreases, and the film becomes less suitable for applications such as lamps. Figure 6.22 shows a typical infrared reflective film produced with these facts in mind. In this diagram, the properties obtained are close to the theoretical calculations.

Table 6.9 shows what happens when these infrared reflecting films are applied to halogen lamps. As shown in Fig. 6.23, by ensuring that the thickness of each film is exactly right for the spectral distribution properties of the filament, infrared rays of about 1 μm are efficiently reflected.

Generally speaking, the efficiency of a lamp is shown by the formula

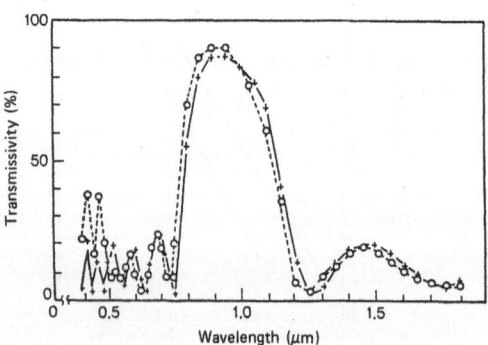

Fig. 6.22 Typical infrared reflection film made with TiO_2 and SiO_2. +, Calculated value; O, actual measurement.

Table 6.9 Properties of lamps coated with infrared reflecting film

Lamp	Power consumed (W)	Power saving (%)	Total luminous flux (lm)	Efficiency (lm/W)
75 W JD 100 V 65 WN-E	65	−13.3	1120	17.2
100 W JD 100 V 85 W N-E	85	−15.0	1600	18.8
150 W JD 100 V 130 WN-E	130	−13.3	2400	18.5

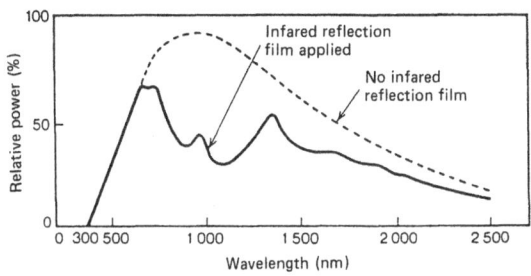

Fig. 6.23 Typical spectral distribution for lamp coated with infrared reflection film.

lm/W, that is, lumen (lm) divided by input power (W). As can be seen from Table 6.9, in this type of lamp, the efficiency is increased by using an infrared reflecting film, and a power saving of about 15% is reached with the same brightness (lumen) as a conventional halogen lamp [21].

6.5 Use of Superfine Particles in High Strength, High Toughness Materials

6.5.1 High Strength Materials

The chief property of ceramics used in high temperature structures – engines, for example – is their high strength at high temperatures. One of the targets for the 21st century is the development of super-strong ceramics, high strength ceramic materials which are even stronger than conventional ceramics.

This aim will be best achieved if we stop confining ourselves to conventional materials such as silicon nitride (Si_3N_4), silicon carbide (SiC) and zirconia (ZrO_2), and consider the possibilities of combining them with many other materials. Of course, the range of materials which can be used is not infinite, and several conditions must be met, especially when the object of the exercise is to produce materials which have high strength at high temperatures.

The production of super-strong end products requires the organic combination of materials selection, mixing, granulation and molding techniques, sintering techniques, machining and joining techniques, and design techniques. In this section, however, we shall focus only on the selection of materials and sintering techniques, from the point of view of the future development of super-hard ceramic materials [22].

The theoretical breaking (tensile) strength σ_{th} of a brittle material is expressed by the following equation [23]:

$$\sigma_{th} = \sqrt{\frac{E\gamma_0}{a_0}}$$

Here, E is the vertical elasticity of the material, γ_0 is the surface free energy, and a_0 is the distance between the atoms.

The best way to increase the σ_{th} of a non-organic material is to choose a material where E and γ_0 are large and a_0 is small. The type of substance which best fulfills these conditions in terms of the bonding energy between atoms and crystal co-ordination is one with covalent bonding, which has an sp^3 hybrid orbit. Next best are metal bonding substances. A compound combining any of the elements in groups I to IV is preferable. Ionic bonds are formed by the large positive and negative ionic forces with the difference in radius, so if the bonding force is on the whole weak, there will be a severe deterioration in strength at high temperatures. Table 6.10 shows the proportions of covalent bonding and ion bonding for typical ceramics, calculated from the difference in electrical negativity between the atoms forming inorganic compounds. The calculation was carried out using the curve proposed by Pauling. This value is fairly approximate, but it is used as a guide to bonding types. As is clear from this table, the covalency ratio is highest for carbides, followed by nitrides and oxides, in that order.

However, those ceramics which can be used in high temperature materials possess all these three bonding types, to some extent, including metallic bonding, and their high temperature properties change according to the proportion of each type of bonding. That is to say, when metal bonding is increased, the number of slip systems increases, and the stacking fault energy grows, so movement gets easier and plastic deformation occurs, and as a result, the material becomes more viscous.

Table 6.10 Ratios for covalent bonding and ionic bonding for typical ceramics

Compound	Diamond	SiC	TiC	ZrC	WC	Si_3N_4	TiN	MgO	Al_2O_3	SiO_2
Difference in electrical negativity	0.00	0.7	1.0	1.1	0.8	1.2	1.5	2.3	2.0	1.7
Proportion of covalent bonding	6.00	0.89	0.77	0.74	0.85	0.70	0.58	0.27	0.37	0.49
Proportion of ionic bonding	0.00	0.11	0.23	0.26	0.15	0.30	0.42	0.73	0.63	0.51

The actual strength of ceramic materials is between 1/100 and 1/1000th of the theoretical strength σ_{th}, which is a long way off the σ_{th}. The main reason for this is that the macroscopic faults (pores, coarse grains, cracks and so on) in ceramic materials cause stress concentration and lead to crumbling. Therefore, when making high strength ceramics, the occurrence of macroscopic faults must be prevented (strength enhancement) and the propagation of cracks must be controlled or prevented (toughness enhancement).

The deflective strength, which is one of the strengths of a ceramic material, can be enhanced in the following ways: (a) by reducing the size of the constituent particles, or (b) by raising the yield limit of the material. Generally speaking, the finer the texture of a polycrystal ceramic material, the greater its strength will be. In other words, the breaking stress can be given by the following equation:

$$\sigma_c = \sigma_0 + \frac{K_c}{\sqrt{d}}$$

Here, σ_0 and K_c are constants, and d is the crystal diameter. The method generally used to make fine ceramic materials is to use superfine particles as the starting material, and sinter them at as low a temperature as possible.

Research based on the considerations mentioned above is being carried out into high strength materials using superfine particles of SiC and Si_3N_4 as the starting materials. First, we shall describe the superfine SiC particles formed by the reactive gas evaporation method [24]. The average diameter of the superfine particles obtained by this method is between 10 and 50 nm, and the bulk density is several percent of the theoretical density. Chemical analysis showed that besides SiC, the particles contained about 10% non-crystalline carbon.

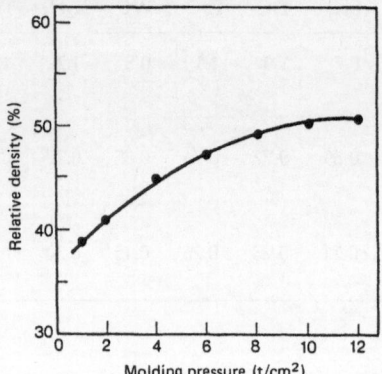

Fig. 6.24 Relationship between uniaxial molding pressure and relative density for superfine SiC powder.

Figure 6.24 shows the relationship between the molding pressure and relative density when superfine SiC particles with a diameter of 25 nm are uniaxially compressed using an 8 mm diameter piston and cylinder. It is known that, when the molding pressure is 10 t/cm², saturation is approached at 50% relative density. Figure 6.25 shows the relationship between the granulity of the powder used as the raw material and the

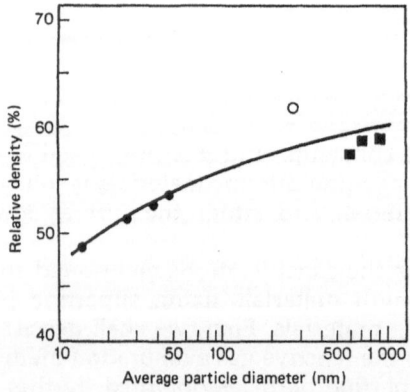

Fig. 6.25 Relationship between particle diameter and relative density for superfine SiC powder. ●, Superfine SiC particles formed by gas evaporation; ○, SiC powder produced by chemical reaction; ■, SiC abrasive (#8000 to 10 000).

relative density of the molding. It can be seen from this diagram that the relative density of the molding decreases as the particles become progressively finer.

To obtain a fine sintered body of SiC, a high relative density is preferable, but when the raw material used is, say, SiC, which is hard and does not easily undergo plastic deformation, it is difficult to raise the relative density above 80%. After this, the density can only be raised by sintering. Sintering is generally promoted by comminuting the raw material powder, and using suitable sintering promoters. It has been confirmed that even when the molding of SiC superfine powder produced by the gas evaporation method is heated to 2000°C or over in argon gas, the relative density barely alters. Therefore, the sinterability must be raised by means such as reducing the particle size still further, eliminating adulteration of the unreacted carbon or Si, and preventing oxide absorption at the surface of the particles [25].

Si_3N_4 is a non-oxide substance, artificially synthesized. It has high covalent bonding and a low self-diffusion coefficient, so it does not sinter easily. Therefore, sintering promoters such as Y_2O_3, SiO_2 and MgO, which produce a liquid phase at high sintering temperatures, are added, and a sintered body is obtained by promoting diffusion at the particle surface. In order to obtain a high strength sintered body of Si_3N_4, the Si_3N_4 powder used must have the following properties: (a) high purity; (b) a narrow granularity distribution, with a particle diameter of 1 μm or less; (c) secondary particles of small diameter; (d) spheroidal particles; and (e) α-type particle shape.

Superfine Si_3N_4 particles can be obtained by various production methods which fulfill these requirements. One of these is the gas phase reaction method, wherein silane or silicon tetrachloride is made to react with ammonia at between 1200°C and 1400°C. These reactions are described by the following formulae

$$3\,SiCl_4 + 16\,NH_3 \rightarrow Si_3N_4 + 12\,NH_4Cl$$
$$3\,SiH_4 + 4\,NH_3 \rightarrow Si_3N_4 + 12\,H_2$$

Highly pure superfine particles with a diameter of between 0.01 and 0.1 μm can be obtained by this method. However, because the amount of powder used is small in proportion to the equipment and energy used, the cost is high. At present, this method is still at the exploratory stage [26].

Sintered bodies of SiC and Si_3N_4, made using superfine particles, are still at the research stage, but, as shown in Table 6.11, these ceramics show promise for practical applications. Figure 6.26 shows a design for a ceramic diesel engine. This is a revolutionary design in which a ceramic material is used for the principal parts of the engine, making use of the heat-resistance and heat-insulating properties of this material, and doing away with the need for the water-cooling system which is the major source of heat-loss, thus allowing large fuel savings.

Table 6.11 Applications of SiC, Si_3N_4 ceramics

Application	Components
Gas turbine	Blades, rotor, stator, nozzles, shrouds Main bearings
Diesel engines	Piston, cylinder, nozzles, tappets, bulb, injectors, turbo-charger, rotor Bearings
Rotating machinery	Ball and rotor bearings
Paper-making machinery	Forming board, deflector, section box cover, wet box cover, felt box
Gas igniters	Igniter components
Other	Radomes, IR windows (IR-radar window) Pump components

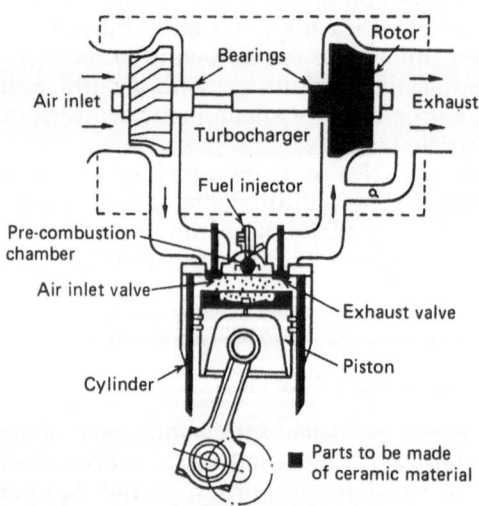

Fig. 6.26 Design for a diesel engine with ceramic parts.

6.5.2 High Toughness Materials

Superfine particles are used in the production of ceramic materials so that their surface energy can be used to promote sintering. The main sintering technique for highly pure raw materials is high temperature sintering, but techniques have recently been developed for reducing the starting materials to superfine particle size, and advances have been made in low temperature sintering using superfine powder. The purpose of the synthesized materials is as follows: (a) to remove impurities; (b) to homogenize the multi-constituent composition; and (c) to comminute and control the particles so that they will sinter more easily.

Among the techniques for producing high performance ceramic materials like these, especially tetragonal-system high toughness zirconia sintered bodies including 2 to 3 mol% Y_2O_3, high density particle materials are outstandingly useful. Attention is now focused on techniques for producing highly pure superfine particles, and these techniques and their industrial applications are an area of great interest.

Zirconia (ZrO_2) is generally known as a ceramic material which sinters with difficulty, in comparison with alumina (Al_2O_3). This is because zirconia has a high sintering temperature, and what is worse, once sintered, it cracks at low temperatures. The reason for this is zirconia's crystal transformation. In other words, zirconia has three crystal systems, as shown in Table 6.12: monoclinic, tetragonal and cubic. The stable phase for monoclinic systems is at 1170°C and below; for tetragonal systems, the temperature range is between 1170 and 2370°C, and for cubic systems, 2370 and 2680°C. Therefore, when ZrO_2 sinters in the cubic system temperature range, after sintering, the crystal system changes from cubic to tetragonal, and as the temperature falls further, the tetragonal system changes to a monoclinic system. This crystal system transformation from tetragonal to monoclinic is termed a martensite transformation, and tends to accompany an increase in volume. The volume increases by about 4.6%, so that many cracks appear in the sintered body, and this can sometimes result in splintering, depending on the circumstances.

In the past, the drawbacks listed above were circumvented using a group of oxides known as stabilizers, such as CaO, MgO and Y_2O_3, which were added in generous quantities to broaden the temperature range over which stable cubic crystals are formed, and to preserve the

Table 6.12

Table 6.12 Crystal systems in zirconia

Crystal system	Monoclinic system	Tetragonal system	Cubic system
	Space group $P2_1/c$ $a = 5.156$ Å $b = 5.191$ Å $c = 5.304$ Å $\beta = 98.9°$	Space group $P4_2/nmc$ $a_1 = 5.094$ Å $c_1 = 5.177$ Å	Space group F_{m3m} $a_c = 5.124$ Å
Density	5830 kgm^{-1}	6100 (Calculated value)	6090 (Calculated value)

cubic crystal structure down to room temperature. Zirconia in which the cubic crystal structure has been stabilized in this way is termed FSZ (Fully Stabilized Zirconia).

Besides FSZ, PSZ (Partially Stabilized Zirconia) has also come in for a good deal of attention recently. In PSZ, smaller quantities of additives are used, so some of the crystals are tetragonal and some cubic, and sometimes a small quantity of monoclinic crystals are also present. If, say, Y_2O_3 is used as the additive, PSZ can be obtained with 2–8 mol% additive, and FSZ can be obtained with 8–12 mol% additive.

When ZrO_2 is fired and cooled, a martensite transformation takes place: the crystal structure changes from tetragonal to monoclinic. The transformation temperature is about 1000°C in the case of simple ZrO_2, and about 350°C in the case of ZrO_2 with 5 mol% Y_2O_3 additive. However, if the ZrO_2 particles are below the critical size, a martensite deformation does not occur, and a metastable tetragonal crystal structure is maintained. The critical particle diameter varies according to factors such as the type of stabilizer, the amount of additive and the raw materials used, but in PSZ containing say, 3 mol% Y_2O_3, it seems to be about 1 μm, and can be achieved by selecting the right firing conditions. Zirconia which contains metastable tetragonal crystals has high toughness, so it is known as TZP (Toughened Zirconia Polycrystalline) in order to distinguish it from ordinary PSZ [27].

At room temperature, the tetragonal crystals in TZP are not completely stable in thermodynamic terms, and exist in a kind of metastable state. This means that when an external force is applied, at room temperature, the crystal structure changes to an even more stable monoclinic structure. The point to note here is that this is a martensite deformation, occurring by a non-diffusion process in which the atoms merely shift out of place. Because this reaction occurs instantaneously, as shown in Fig. 6.27, the impact of the external force is absorbed and provides the energy for the transformation. Moreover, because the volume increases, it is an extremely effective way to reduce the occurrence of cracks which lead to crumbling.

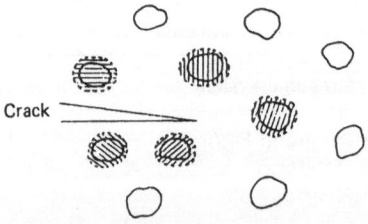

○ Metastable zirconia particles (monoclinic crystals)

◈ Zirconia particles (monoclinic crystals) produced in transformed martensite

Fig. 6.27 Simplified illustration of martensite transformation.

The phenomenon of transformation caused by stress is called Stress Induced Transformation. This is what distinguishes the toughness of zirconia from that of other ceramics.

The high toughness of TZP can be shown using the following parameters. The stress-intensity factor K_1 shows the stress at the tip of the crack and the strain state. K_{1c} is the critical value at which cracks grow rapidly. The energy needed for destruction is proportional to the square of K_{1c}. The greater the value for K_{1c}, the harder the material is to break, and the higher the toughness. The K_{1c} for TZP, as shown in Table 6.13, is between 9 and 10 (units: $MN/m^{3/2}$). The value for Si_3N_4 is between 4 and 6, the value for SiC is up to 3, and the value for Al_2P_3 is 3: this shows how tough TZP is [28].

When synthesizing TZP, it is important to reduce the particle size of the sintered body. In ceramics obtained from superfine particle materials by low temperature sintering, abnormal particle growth tends not to occur, so these materials are suitable for applications which require small, evenly-sized particles. Low temperature sintering processes using alkoxides as the raw materials have good potential for use in the production of high toughness zirconia.

Table 6.14 lists sintering densities obtained by several different processes [29]. Of particular interest is the report [30] describing how a sample with a relative density of 99% was obtained at 1100°C – item 1 in this table. One method used for adjusting superfine powder materials is shown in Fig. 6.28 [31]. In this process, flocculated particles are removed by filtering or other methods, and the superfine particles ($\leqslant 0.01\ \mu m$) remaining in the solution are collected by the centrifugal precipitation method, molded, and then sintered. Sample 3 in Table 6.14, which still contained flocculated particles when it was sintered, has a high sintering temperature: 1450°C. This is thought to be due to the poor packing of the powder inside the molding, which in turn is due to the flocculated particles.

Table 6.13 Breaking force (K_{1c}) of ·various ceramics

Material	K_{1c} ($MN/m^{3/2}$)
ZrO_2–Y_2O_3	6–9
ZrO_2–CaO	9.6
ZrO_2–MgO	5.7
ZrO_2	1.1
Al_2O_3–ZrO_2	9.8
Si_3N_4	4.8–5.8
SiC	3.4
B_4C	6.0
Al_2O_3	4.5
Monocrystalline spinel	1.3

Table 6.14 Sintering by various processes

No.	Zirconia superfine powder	Synthesis method	Firing temperature (°C)	Sintering density (%)
1	6 mol% Y_2O_3	Alkoxide method	1100	99
2	6.5 mol% Y_2O_3	Neutralization and co-precipitation method	1300	98
3	6 mol% Y_2O_3	Alkoxide method	1450	98
4	16 mol% CaO	Freeze-drying method	1500	98
5	3 mol% Y_2O_3	Neutralization and co-precipitation method	1500	>99
6	4.5 mol% Y_2O_3	Neutralization and co-precipitation method (washing in water)	1500	99.5

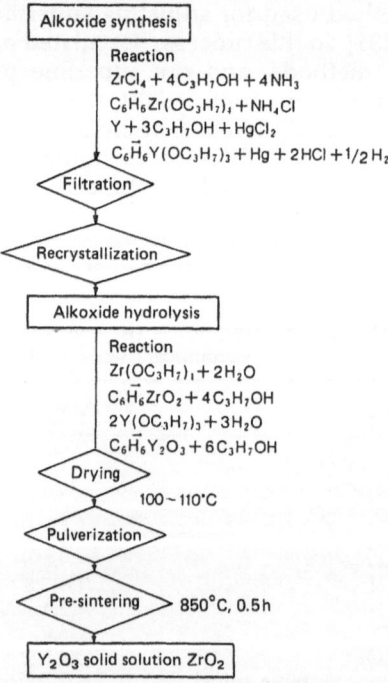

Fig. 6.28 Production process for Y_2O_3–PSZ using the alkoxide method.

We shall now consider how the powder characteristics of superfine powders affect sintering. The flocculated particle model shown in Fig. 6.29 is helpful here [32]. First, the crystals in this diagram are in the single crystal state, and judging from the broadening of the X-ray analysis peak, many of the particles have a diameter matching with BET. Another assessment method which has become available more recently is the lattice image provided by TEM photographs. It increases in size as the presintering temperature rises, and could also be seen as an index to the "activity" of the particles.

Although in some cases the primary particles are crystals, it is better to think of them as distinct from crystals. They can be measured using an SEM and other types of microscope. The particle diameter can be controlled by regulating the nucleus formation speed and the particle growth speed during the reaction of precipitation from solution (including the alkoxide hydrolysis reaction). By optimizing the reaction conditions, it is even possible to regulate the particle size and thereby obtain the ideal monodispersed spheroidal fine particles. In the case of low temperature sintering, this is an extremely important factor.

Although the particles are flocculated, the degree of flocculation can be altered by altering the diameter of the primary particles. This can be controlled to some extent by mechanical methods such as ball milling. If the flocculated particles are not hard, there is no problem, and there are various ways and means of obtaining soft flocculated particles. The degree of flocculation is affected by the "packability" of the powder in the molding, which is important in low temperature sintering.

The shape and diameter of the primary particles – the critical factor in low temperature sintering – can be controlled by selecting the conditions for the alkoxide hydrolysis (type of alkoxide, solvent, concentration, temperature, speed of hydrolysis and other parameters).

The two kinds of 3 mol% Y_2O_3–ZrO_2 powder shown in Table 6.15 were

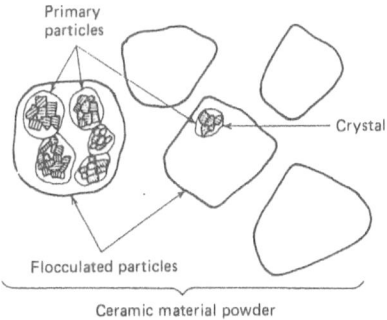

Fig. 6.29 Flocculated particle model.

Table 6.15 Powder characteristics of 3 mol% Y_2O_3 PSZ obtained by the alkoxide method

Sample	Pre-sintering temperature (°C)	Crystal diameter[a] (nm)	Primary article diameter[b] (nm)	Flocculated particle diameter[c] (μm)
0.03 μm	800	12	30	2.8
0.10 μm	800	12	100	1.7
Commercially available product obtained by the alkoxide method	?	14	<50	2.2

[a]X-ray analysis.
[b]SEM.
[c]Granularity distribution using centrifugal precipitation measured by phototransmission.

adjusted and the relationship between sintering temperature and sintering density was found to be as shown in Fig. 6.30 [33]. For comparison, the diagram also shows a commercially available 3 mol% Y_2O_3–ZrO_2 obtained by the alkoxide method. These types of powder can be classed as low temperature sintering powder and high temperature sintering powder, according to the diameter of the primary particles. In the low temperature sintering type, a relative density of 99% was obtained at 1300°C, and in the high temperature sintering type, almost no abnormal grain growth was observed even after sintering at 1600°C.

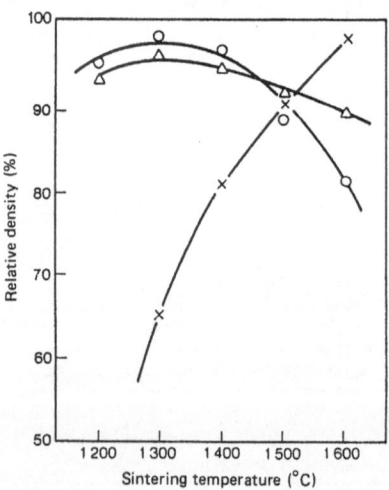

Fig. 6.30 Sintering characteristics of 3 mol% Y_2O_3–PSZ obtained by the alkoxide method. △, Commercially available product; ○, 0.03 μm; ×, 0.10 μm.

New applications for TZP made from superfine particle materials have only just begun to be developed. The main applications are in the field of edged tools such as kitchen knives and industrial cutters. It is also used in fields such as wear-resistant components like ball mills, dies for wire-drawing, and cutting tools. Zirconia ball mills are suitable for use in the production of high purity powders where abraded dust is particularly undesirable, such as electronic components and magnetic materials. When the zirconia balls used are between 2 and 3 mm in size, the particles can be comminuted to superfine particle size [34]: TZP will probably continue to be used in the future for superfine particle production.

TZP also has a high breaking toughness, as well as heat insulating properties, so it has great potential for use in heat insulating diesel engine components.

6.6 Use of Superfine Particles in Catalyst Materials

Two types of superfine metal particles are used as catalyst materials. The first type are superfine metal particle homogenous catalysts, wherein metal particles with a diameter of up to several tens of nanometers are suspended just as they are in a liquid phase medium, or used on a carrier as required. The second type are carrier metal catalysts, which are regulated such that metal particles with a diameter of between 1 and 10 nm disperse well on the surface of oxides. These are mainly used in solid–gas non-homogenous systems. We shall deal with the superfine metal particle homogenous catalysts first [35].

6.6.1 Metal Superfine Particle Homogenous Catalysts

A catalyst material is a functional element which steadily promotes a chemical reaction when supplied with heat. Therefore, the catalyst's activity (the extent to which the catalyst's active sites accept the reagents), selectivity (the ability to distinguish between substances), durability (the extent to which the catalyst promotes the desired reaction), and performance (efficiency of operation) should be as high as possible. These are all vital factors in a catalyst.

The activity of a solid catalyst is proportional to the specific surface area of the active component. In metal superfine particles, the smaller the particle diameter, the greater the specific surface area and the greater the number of active sites per unit weight of catalyst. Moreover, the smaller the particle diameter, the greater the irregularity of the surface, so when the irregularities are the parts that carry out the catalyst function, the reaction speed is higher per active site. The smaller the particle diameter, the greater the amount of catalyst which can be stuffed into the reaction container, and the higher the space–time yield rate of the reaction device.

Two of the properties of solids are structural susceptibility and structural non-susceptibility. The catalytic reactivity of a solid is an important structural susceptibility property. In the field of catalysts, reactions which are susceptible to the relative activity of the catalyst or to the particle diameter are classed as structural susceptibility reactions, and reactions which are not susceptible to such things are classed as structural non-susceptibility reactions [36]. Superfine particles are used as catalysts precisely because they have good potential for structural susceptibility reactions. As shown in Fig. 6.31, there are four types of structural susceptibility reactions: those in which the relative activity increases as the particle diameter decreases, those in which the relative activity decreases as the particle diameter decreases, those which show a peak value at a certain particle size, and those in which the particle size has almost no effect.

Here is an example of a reaction in which the relative activity increases as the particle diameter decreases: a reaction has been reported [37] in which the activity was seen to increase remarkably as the Pt surface area increased, and the active energy of the reaction was seen to decrease as

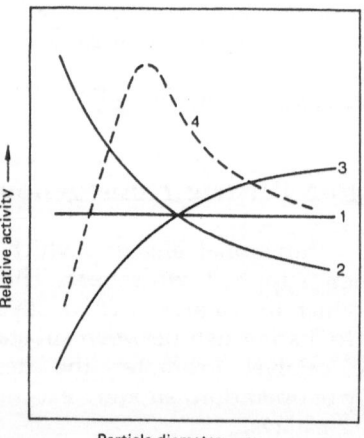

Fig. 6.31 Relationship between particle diameter and relative activity in metal particles. 1, Reaction in which particle size is irrelevant; 2, reaction in which relative activity decreases as particle size increases; 3, reaction in which relative activity increases as particle size increases; 4, reaction which peaks between particle diameter and relative activity.

the relative surface area increased. As the metal particles get smaller, they become more active, so there is probably an activity level at which the properties change.

Here is an example of a reaction in which activity decreases along with particle diameter. A study was carried out [38] on the relationship between particle diameter and propane oxidation in Pt carried on alumina. It was found that when the particle diameter was between 2 and 200 nm, the activity decreased. The reason for this is related to the change in the ratio of adsorbed oxygen in the oxidation of propane, due to the amount of electron migration between the Pt particles and the carrier.

One disadvantage of metal superfine particle homogenous catalysts which is closely related to the smallness of the particle diameter is the fact that high surface reactivity causes a lack of durability. A heat supply is indispensable when promoting a chemical reaction, but heat brings about various surface reactions such as melt-adhesion between the superfine particles, sintering, and reaction with the carrier, leading to changes in characteristics. This can be prevented by various means, such as selecting a good carrier, using mild reaction conditions, using a suspension dispersion state in the case of a liquid phase reaction, or using an endothermic reaction.

Metal superfine particles regulated by a "dry process" such as in-gas vapor deposition, gas phase chemical reaction or the laser method have all been involved in high temperature processing, and if they are used under the right conditions, they can exhibit the durability necessary for catalyst materials. In the category of metal superfine particles obtained by the in-gas vapor deposition method, catalytic active components such as nickel (Ni), cobalt (Co) and iron (Fe) are commercially available in the form of spheroidal particles with a sharp particle diameter distribution, so these are the ones in which the catalytic effect has been most thoroughly tested.

To cite an example, it has been shown [39] that superfine Ni particles with a diameter of 20 nm, regulated by the in-gas vapor deposition method, are an excellent liquid phase dehydration catalyst for 2-propanol. Using gas chromatography on the liquid and gas phase components, it was confirmed that the only substances generated were acetone and hydrogen.

2-propanol dehydration reaction:

$$(CH_3)_2CHOH \rightarrow (CH_3)_2CO + H_2$$

Acetone hydrogenation reaction:

$$(CH_3)_2CO + H_2 \rightarrow (CH_3)_2CHOH$$

In the above reaction, the metal Ni superfine particle homogenous catalyst is placed in the liquid phase suspension reaction container in a high dispersion state, and tends to produce a high space–time yield rate.

Next, let us briefly touch upon the catalytic effect of superfine particles of precious metals. Hirai et al. found that in the presence of a high polymer surface active agent such as polyvinyl alcohol or polyvinyl pyrrolidone, when a precious metal salt was dissolved in an alcohol such as methanol, reflux heated, and reduced by the alcohol, superfine precious metal particles of uniform diameter were generated [40]. The solvent was removed by drying, and when the powder thus obtained was once again dispersed in a solvent, it returned to its original metal colloid state. It seems that when superfine particles are properly enclosed by a high polymer functional group and a hydrophobic group, they produce a colloid which is stable and easy to handle.

Precious metal superfine particle homogenous catalysts regulated in this way show a very high activity and selectivity in olefine oxygenation reactions, for example. The high volume displacement group in the olefine double bond is only effectively oxidized by the adjacent substrate in the presence of a metallic rhodium superfine particle homogenous catalyst with a particle diameter of 1 nm or less. Table 6.16 shows the relationship between particle diameter and oxygenation catalyst activity for a range of olefines.

6.6.2 Carrier Metal Catalysts

In a carrier metal catalyst, the metal particles are fixed on the surface of a carrier (usually an inorganic substance such as silica or alumina) which has a large surface area [41]. The metal particles usually have a diameter of up to several tens of nanometers. Below a certain size limit, metal particles lose their metallic properties. Metal compounds in the form of very small particles also show characteristics which are different from

Table 6.16 Olefine hydrogenation using superfine particles of metallic rhodium as a catalyst

Olefine substrate radical	Catalyst activity[a]		
	Rh-PVP-MeOH/H$_2$O (diameter: 3.4 nm)	Rh-PVP-EtOH (diameter: 2.2 nm)	Rh-PVP-MeOH/NaOH (diameter: 0.9 nm)
1-hexene	15.8	14.5	16.9
Cyclohexene	5.5	10.3	19.2
2-hexene	4.1	9.5	12.8
Methyl vinyl ketone	3.7	4.3	7.9
Methyl oxide	0.6	4.7	31.5
Methyl acrylate	11.2	17.7	20.7
Methyl methacrylate	5.8	15.1	27.6
Cyclo-octene	0.6	1.1	1.2

[a]Speed of hydrogen adsorbtion in methanol: (H$_2$ mol/Rh g.atom.s); 30°C; atmospheric pressure: H$_2$ 1 atm; (polyvinyl pyrrolidone-coated metallic rhodium) = 0.01 m mol/dm^3; (olefine substrate) = 25 m mol/dm^3.

the characteristics of the compound in its bulk state. Since the catalytic effect of a solid reflects its surface properties, the way these properties change when the particles are comminuted is a matter of great interest. Those changes which are relevant to the catalytic effect are as follows: (a) the increase in the number of electrons around the edges and on the top of the fine metal crystal surface; (b) the change in the electronic state which is assumed to occur when the particle diameter is 5 nm or less [42]; and (c) the electronic change and the several chemical changes caused by interaction with the carrier (the carrier effect) [43]. A carrier catalyst of superfine metal particles is likely not only to increase the surface area of the metal particles and heighten the catalytic activity, but to increase the selectivity also, due to the effects listed above. Below, when we have briefly touched on methods for regulating carrier catalyst superfine particles, we shall give examples of the carrier superfine particle homogenous catalysts (b) and (c) above.

Broadly speaking, carrier catalyst superfine particles can be regulated in the following five ways:

1. Impregnation method: when the amount on the carrier is small, superfine particles with a diameter of about 1.5 nm can sometimes be obtained (1% Rh/Al_2O_3) [44].

2. Ion exchange method: positive ions such as H^+ or Na^+ on the carrier surface are exchanged for positive complex ions (such as $Pt(NH_3)_4^{2+}$, $Rh(NH_3)_5Cl^{2+}$), and superfine particles with a diameter of about 1 nm can be carried (Pt, Ru/zeolite, Pt/SiO_2) [45].

3. Carbonyl cluster adsorbtion method: a complex such as $Rh_6(CO)_{16}$ or $Ru_3(CO)_{12}$ is adsorbed onto a carrier from an organic solvent, and when degradation and reduction processing has been carried out, superfine particles with a diameter of about 1 nm are obtained (Rh/SiO_2, Ir/Al_2O_3, Fe/Al_2O_3, Ru/Al_2O_3) [46]. From the mixed cluster, superfine alloy particles are obtained (Co–Rh/Al_2O_3, Ru–Fe/Al_2O_3, Os–Rh/Al_2O_3) [47].

4. Metal evaporation method: a metal is heated and evaporated in the presence of an organic molecule gas, and fixed on a carrier (Co–Mn/SiO_2) [48]. Alternatively, the metal can be fixed directly to the carrier by metal evaporation [49].

5. Alkoxide method: an alkoxide is mixed from a metal ethylene glycol salt and a carrier, then gelated, fired and reduced (Ni/SiO_2, Rh/SiO_2) [42].

We shall now look at examples of carrier superfine particle homogenous catalysts we earlier labeled (b) and (c).

Some superfine particle alloys have been found to show a much higher activity than conventional catalysts. Co–Mn/SiO_2 onto which a vapor of Co and Mn has been directly fixed is highly active in oxygenating ethylene [48]. Pt–Mo/zeolite obtained by reacting $Mo(CO)_6$ with Pt/zeolite shows excellent activity in butane hydrogenolysis [50]. Fe–Ru/Al$_2$O$_3$ obtained from mixed clusters of Fe and Ru is highly active in the homogenation of ethylene [51].

The polyhydric alkoxide method is worthy of attention as a way of carrying transition metal, particularly Ni, on silica or alumina, and regulating metal catalyst fine particles of uniform diameter [52]. In one widely applicable method, a metal salt is heated and dissolved in a polyhydric alcohol such as ethylene glycol. Ethyl silicate, for example, is added to the result and well mixed, more water is added, and the result is stirred; when the gel produced has been dried under reduced pressure, it is fired in air and reduced in a hydrogen current. Table 6.17 shows an example of a catalyst regulation procedure using the alkoxide method. For reference, this table also shows an example using the impregnation method. When metallic nickel particles with a diameter of between 2 and 3 nm are carried on silica, the reaction of propion aldehyde in hydrogen has a markedly better alcohol/CO generation ratio [53]. Ruthenium catalysts regulated by this method also work excellently in selective hydrogenation reactions for obtaining cyclohexane from benzene [54].

$$CH_3CH_2CHO \xrightarrow{H_2} \begin{array}{l} \nearrow CH_3CH_2CH_2OH \\ \searrow CH_3CH_3 + CO \end{array}$$

$$C_6H_6 + 2H_2 \rightarrow C_6H_{10}$$

Finally, we shall look at an example of the improvement in selectivity

Table 6.17 Catalyst regulation procedures using the alkoxide method and the impregnation method

	Alkoxide method	Impregnation method
1	Nickel nitrate is dissolved in ethylene glycol at 80°C	Silica powder is dipped in an aqueous solution of nickel nitrate
2	Ethyl silicate is added, and the result is mixed and stirred at 80°C	The result is heated in a water bath and stirred continuously
3	Water is added, and the stirring is continued, whereupon a gel is formed. The gel is dried under reduced pressure and then pulverized	The heating and stirring are continued and the water is evaporated, leaving a powder
4	The powder is dried in a drier	The powder is dried in a drier
5	The powder is fired in air for 4 hours at 450°C	The powder is fired in air for 4 hours at 450°C
6	The result is reduced in a stream of hydrogen for 4 hours at 450°C	The result is reduced in a stream of hydrogen for 4 hours at 450°C

brought about by comminution to superfine particle size. It has been reported [55] that when toluene was hydrogenated using a catalyst obtained by the metal evaporation method and the impregnation method, the former was found to have the higher selectivity. It has also been shown [56] that, in the hydrogenation of propion aldehyde, Ni/SiO_2 with a particle diameter of 5 nm is highly selective. The addition of K to high-dispersion Ru/Al_2O_3 regulated from $Ru_3(CO)_{12}$ greatly improves the proportion of olefine in the $CO + H_2$ reaction [57].

When CO hydrogenation is carried out with Fe/Al_2O_3 (Fe particle diameter: 2 nm) obtained from $Fe_3(CO)_{12}$, propylene is obtained with a selectivity rate of 45%. However, the Fe flocculates (forming particles 50 nm in diameter) during the reaction, and the selectivity rate falls [58]. When CO hydrogenation is carried out using a catalyst obtained by high-dispersion of Co or Fe on a zeolite carrier, the selectivity of the propylene or butene is high, although the low conversion rate is a problem.

Catalysts obtained by fixing metal particles with a diameter of between 1 and 10 nm on a porous carrier such as alumina, silica, magnesia, titania or zeolite are widely used in factories. The regulation of such catalysts is an extremely important matter, which affects their activity, selectivity, life, and other factors. Sintering is an unavoidable problem in all carrier metal catalysts. This is the phenomenon whereby the activity drops during the reaction although absolutely no impure substance is included. In other words, the metal particles fuse, forming even larger metal particles, and the exposed surface of the metal particles becomes smaller, so the active sites decrease, and the phenomenon of ageing occurs. This phenomenon has been investigated by many researchers, and the separation, re-dispersion and fusion of metal particles have been directly and indirectly observed. As shown in Fig. 6.32, there are several models for the particle growth mechanism, but the question of how to regulate

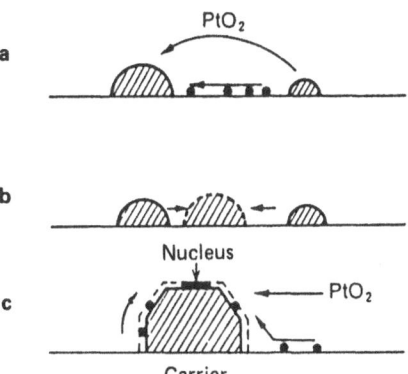

Fig. 6.32 Models of the particle growth mechanism. ●, Platinum atom; ▨, platinum particle. a Model proposed by Flynn and Wanke, b model proposed by Ruckenstein and Pulvermacher, and c model proposed by Wynblatt and Gjostein.

catalysts with high activity and long life – in other words, catalysts which are resistant to sintering – is the greatest obstacle to be overcome in the practical application of carrier metal catalysts.

6.7 Use of Superfine Particles in Sensor Materials

6.7.1 Optical Sensor Materials

Sensors consisting of superfine particles have the following characteristics: they have a large specific surface area, and the larger areas of contact between the superfine particles mean larger changes in electrical resistance with the adsorption and de-adsorption of gas. In such sensors, the gas sensitivity can be heightened and the detection range broadened at the same time, and the functional requirements such as response speed and detection range are fulfilled. The structure of the sensing film can be controlled by controlling the production conditions, so it is likely that, by selecting the working temperature and other factors, the selectivity can be improved. That is to say, although they are less accurate than other semiconductor sensors, they have good potential in terms of other functions.

Gas sensors are typical of chemical sensors, so we shall deal with these first. A gas sensor detects and measures gas, using the characteristics of semiconductors – especially the electrical properties of metal oxides – which vary according to the composition of the gases in the atmosphere. Since the report by Seiyama et al. [59] in 1962, research and development has been carried out on sensor materials vis-à-vis various gases [60]. Combustible gas leak alarms using SnO_2 are already in practical use.

The semiconductor particles used in sensors lie in a wide size range between several nanometers and several micrometers. However, since the vital factor is the interaction between the area around the surface of the particles and the atmospheric gas, the finer the particles and the greater the specific surface area, the more effective the sensor will be. Also, when the particles are comminuted to superfine size, their properties (that is, their conduction, gas adsorption and oxide reduction properties) change, and in many cases this produces new sensor functions.

We shall now take ferric oxide (Fe_2O_3) as an example of a substance which can be used as a gas sensor when in the form of superfine particles triferrous tetroxide [61]. Generally speaking, there are three iron oxides – iron(II) oxide (FeO), iron(III) oxide (Fe_2O_3), and (Fe_3O_4) – but the latter two are the most important in terms of electronic materials. Iron oxides have conventionally been used in a wide range of applications such as ferrite raw materials, magnetic recording materials, and cosmetics. As

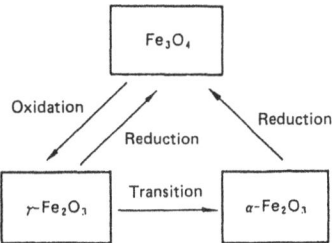

Fig. 6.33 Iron oxide oxidation, reduction and transition processes.

gas sensors, however, they have little sensitivity, so they have not attracted much attention in this area. There are two types of Fe_2O_3: α-Fe_2O_3, which has a corundum-type crystal structure, and γ-Fe_2O_3, which has a spinel-type crystal structure. As shown in Fig. 6.33, conversion between these two substances and magnetite – spinel-type crystal structure Fe_3O_4 – is carried out by the processes of oxidation, reduction and transition. A reversible oxidation–reduction process can be used to convert between Fe_3O_4 and γ-Fe_2O_3. The α and γ phases of Fe_2O_3, however, have very different crystal structure and fundamentally different electrical properties.

Figure 6.34 shows the working temperature properties of sensitivity to combustible gases such as isobutane (i-C_4H_{10}). When γ-Fe_2O_3 is heat-treated at a high temperature and made to undergo the transition to α-Fe_2O_3, it appears to lose almost all its gas sensitivity. This is probably the reason why Fe_2O_3 has never attracted much attention as a gas sensor.

However, as shown in Fig. 6.33, since α-Fe_2O_3 can be converted to

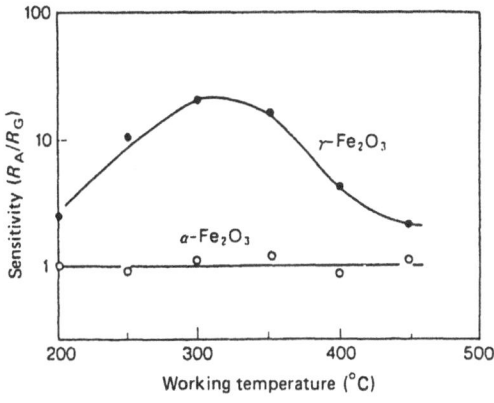

Fig. 6.34 Relationship between sensitivity and working temperature.

Fe_3O_4 by reduction, it ought to be possible to make it gas-reactive by increasing the activity of the sensor in some way. Aiming to obtain α-Fe_2O_3 with high activity regarding reductive gases, Nakatani et al. [62] succeeded in obtaining gas sensor materials from various iron salts using the wet method, regulating α-Fe_2O_3 in the superfine particle form shown in Table 6.18. Some of the items in this table were made from starting materials consisting of natural mineral salt containing sulfuric acid ions. These items contain minute amounts of sulfuric acid ions, and this fact plays a fundamental role in the comminution of the sensor to superfine particle size, the conversion to a non-crystalline substance, and thus in the achievement of gas sensitivity.

Figure 6.35 shows the effect of additives on gas sensitivity (R_A/R_G, working temperature 400°C). For reference, this diagram also shows an α-Fe_2O_3 sintered body produced using iron(III) nitrate, as an example of an iron salt which does not contain any sulfuric acid ions. From this diagram it can be seen that gas sensitivity is produced by using a metal salt containing sulfuric acid ions, and markedly increased by the additives Ti^{4+}, Zr^{4+} and Sn^{4+}. When the sensor is produced by adding Sn^{4+}, the particles obtained by comminuting this sensor are particularly fine, their average diameter being 10 nm, as can be observed using a transmitting electron microscope. As a result, they have a very large specific surface area – 125 m^2/g –which leads to improved gas sensitivity. The magnitude of the effect can be judged from the fact that in the case where Sn^{4+} was not added, the average particle diameter is 50 nm and the specific surface area is 32 m^2/g.

Considering the similarity between IC integration techniques and techniques for producing superfine particle films, and the extent to which they are interrelated, it seems likely that there will be a large demand in future for the development of sensors integrated with IC chips. There is a high possibility that superfine particle technology will be used to produce highly responsive, miniaturized sensors with low energy consumption and a wide range of functions.

The technology is now moving away from the application of superfine particles in "superfine particle films" in the form of a film on a substrate, and towards superfine particle sensors [63]. In other words, "superfine particle gas sensitive films" with a special film structure in which superfine oxide particles, with an average diameter in a fixed range between one and several nanometers, are aligned in a specific direction, will be used to make gas and humidity sensors with special properties lacking in conventional sensors.

When forming superfine particle films for use in sensors, it is important to control the constitution and average diameter of the superfine particles. Tin oxide (SnO_2) has been investigated as an example of an oxide composition: the raw material is heated and evaporated in gas under reduced pressure, then the vapor is cooled in gas, whereupon it condenses to produce superfine SnO_2 particles [64]. The structure of the film, consisting of porous, columnar superfine SnO_2 particles produced in

Table 6.18 Effect of starting materials and additives on gas sensitivity, specific surface area, and average crystal size of the α-Fe_2O_3 sintered body

Sample	Sensor		R_A (kΩ)	Sensitivity R_A/R_G (0.5)					Specific surface area (m²/g)	Average crystal nucleus size (nm)
	Starting material	Additives[a]		CH_4	C_3H_8	1-C_4H_{10}	H_2	C_2H_5OH		
A	$FeCl_3$–$6H_2O$	—	78	0.86	0.42	0.40	0.57	1.0	12	200
B	$Fe(NO_3)_3$–$9H_2O$	—	502	0.97	1.1	1.2	0.98	11	11	100
C	$FeSO_4$–$7H_2O$	—	1300	3.1	12	14	6.4	8.7	31	60
D	$Fe_2(SO_4)_3(NH_4)_2$ SO_4–$24H_2O$	—	4050	3.0	11	15	6.8	8.8	38	40
E	$Fe_2(SO_4)_3$–nH_2O	—	800	2.5	11	14	5.5	7.4	32	50
F		$Ti(SO_4)_2$ sol	230	5.0	34	42	13	11	66	40
G		$ZrOCl_2$–$8H_2O$	1700	4.6	36	48	14	9.2	86	30
H		$SnCl_4$–$5H_2O$	430	6.6	25	29	12	12	125	10
I	Fe_2O_3	—	1260	1.0	1.7	1.8	1.2	1.4	10	10 >200

[a]Additives were added in 20 mol% in the form of TiO_2, ZrO_2 and SnO_2.

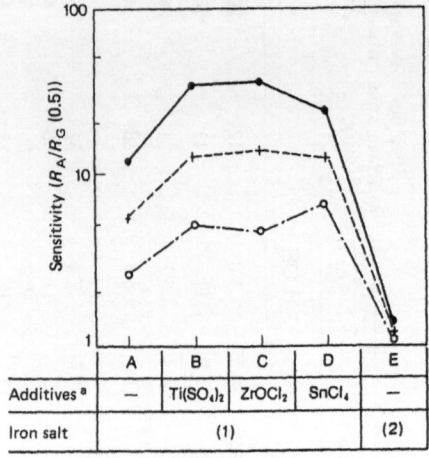

	A	B	C	D	E
Additives [a]	—	Ti(SO$_4$)$_2$	ZrOCl$_2$	SnCl$_4$	—
Iron salt		(1)			(2)

(1) : Fe$_2$(SO$_4$)$_3$-nH$_2$O
(2) : Fe(NO$_3$)$_3$-9H$_2$O
[a] Amount of additive: 20 mol%

Fig. 6.35 Relationship between starting material salt, additives, and gas sensitivity in an α-Fe$_2$O$_3$ sintered body. ●, C$_3$H$_8$; +, H$_2$; O, CH$_4$.

oxygen under a pressure of 0.5 Torr, is shown in Fig. 6.36. The film's electrical conductivity is high in the direction perpendicular to the substrate, because the superfine SnO$_2$ particles consist of densely packed cylindrical columns, but very low in the direction parallel to the substrate, because of control into long thin channel areas. When the film is used as a multi-function sensor to measure factors such as gas and humidity, the channels which link together the columns in Fig. 6.36 play a major

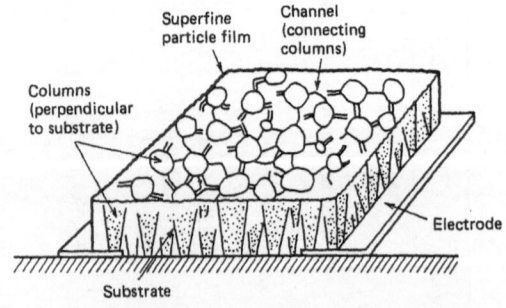

Fig. 6.36 Structure of superfine particle film.

role. As shown in Fig. 6.37, this configuration has been found to possess ideal properties for use as gas/humidity sensors wherein the electrical conductivity of the film decreases rapidly when the average particle diameter is no more than 6 nm [65].

As shown in Fig. 6.38, a gas-sensitive film made from superfine particles can be made to detect selectively several different gases by altering its working temperature. It has been shown that a film formed in oxygen gas at 0.5 Torr, can be made to detect selectively H_2O, C_2H_5OH and iso-C_4H_{10} by changing the working temperature of the film from low to high.

The reason for the high sensitivity of sensors using a gas-sensitive film made from superfine particles is as follows: in conventional thin films, the mobility does not vary according to the gas detected, but in a film made of superfine particles, as shown in Fig. 6.39, the carrier concentration n varies, and so does μ.

As shown in Fig. 6.40, an integrated superfine particle gas sensor [66] is a device made by integrating a gas-sensitive film made of superfine particles with an IC chip. The functional characteristics of such a sensor can be summarized as follows:

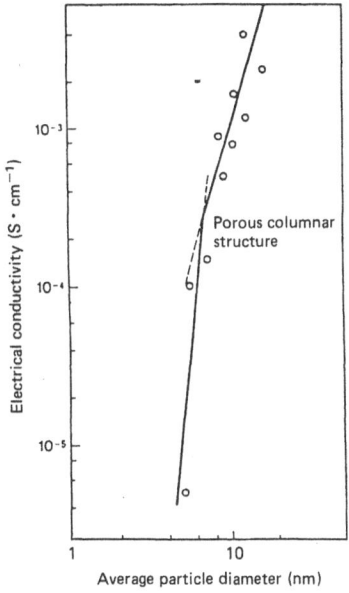

Fig. 6.37 Relationship between electrical conductivity and average particle diameter in a gas-sensitive film made from superfine particles of porous oxide.

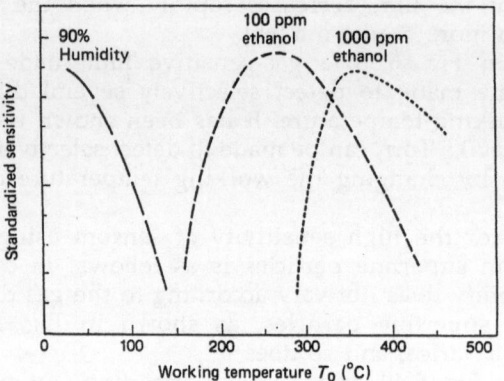

Fig. 6.38 Relationship between gas-sensitivity and working temperature in a gas-sensitive film made from superfine particles.

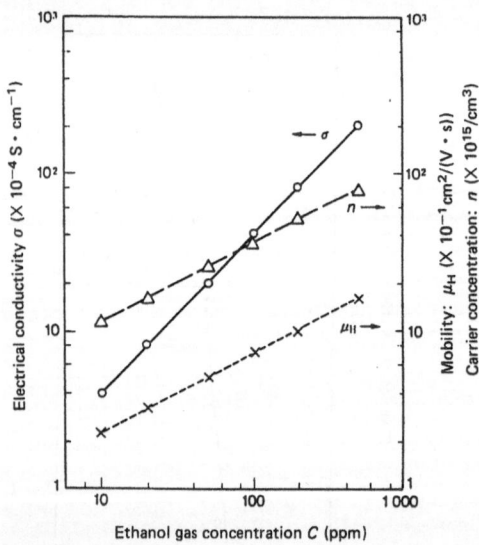

Fig. 6.39 Relationship between ethanol gas concentration, electrical conductivity σ, rate of mobility μ_H, and carrier concentration n in an atmosphere of air containing ethanol.

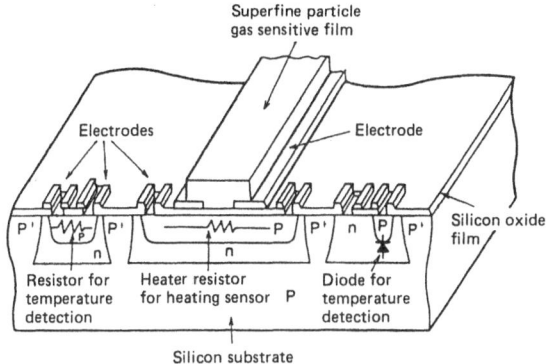

Fig. 6.40 Structure of integrated superfine particle multi-function sensor.

1. It is a "composite sensor" combining a gas sensor, humidity sensor and temperature sensor on the same IC chip, and has a wide variety of measurement functions.
2. It is a "multi-function sensor", wherein one superfine particle gas detection film can be made selectively to detect several different gases by altering the working temperature.
3. Because the particles used have an average diameter of several nanometers the specific surface area is extremely large – several hundred m²/g. As shown in Fig. 6.36, the special composition of the film results in high sensitivity and a high response speed.
4. Because the superfine particle gas-sensitive film is integrated with the film heater and the temperature sensor, the working temperature of the film can be controlled accurately.
5. Because the sensor chip is very small – 1–2 mm square – and because the film has a small heat capacity, its power consumption is low.

The weight of oxide particles which make up the superfine particle gas-sensitive film need only be very small – about 0.5 μg. Barely a gram of raw material is needed to make 2 million elements. Combining low power consumption and small chip size, these sensors could be called "resource-saving and energy-saving sensors". Sensors using superfine particles are well suited to the times, and have great potential for future development.

6.7.2 Infrared Sensor Materials

Optical sensors can be obtained by making films from superfine particles of metals such as gold. Figure 6.41 shows the relationship between optical

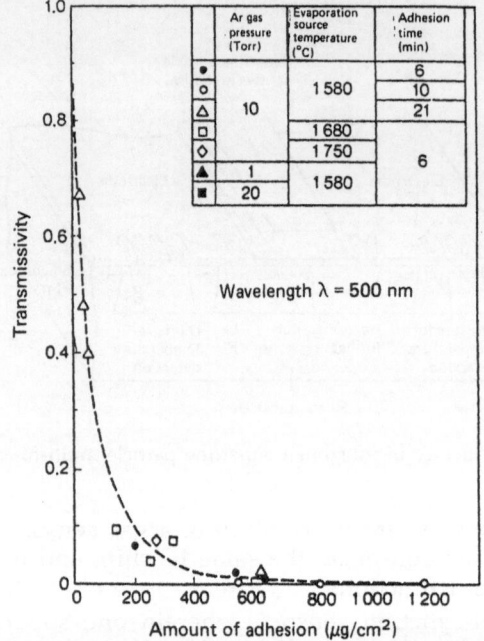

The table shown in the figure:

	Ar gas pressure (Torr)	Evaporation source temperature (°C)	Adhesion time (min)
●		1 580	6
○	10		10
△			21
□		1 680	
◇		1 750	6
▲	1	1 580	
■	20		

Wavelength λ = 500 nm

Fig. 6.41 Relationship between optical transmissivity and amount of adhesion per unit surface area in a film made from superfine particles of gold.

transmissivity and film thickness when a film made from superfine gold particles is formed on a glass plate by in-gas evaporation [67]. Just by controlling the amount of adhesion per unit surface area, it is possible to obtain a fixed optical transmissivity which is independent of Ar gas pressure, evaporation source temperature, adhesion time, and other factors. The optical reflectivity is also very small, and when the superfine particle film is formed with 500 μg/cm^2 or over, a high light absorption rate – 95% or over – can be obtained.

The characteristics of a film made from superfine particles of gold are that it has a high light absorption rate, and the light absorption rate ranging from the visible spectrum to the infrared area is largely independent of wavelength. These properties can be used in infrared sensors.

Figures 6.42 and 6.43 show detailed structural diagrams [66]. In the sensor in Fig. 6.42, a thin film (with layers of Si$_3$N$_4$, SiO$_2$ and Si$_3$N$_4$) is formed on one part of the 4 × 4 mm^2 element, and a 1 × 1 mm^2 film made of superfine gold particles is formed on top of this. In between, a thermocouple consisting of a thin film of InSb and Te is connected in series. The thickness of the silicon substrate is 200 μm, and since silicon's heat conductivity is 100 times greater than that of the laminated section, a cold junction can be set up on the thin film on the silicon substrate.

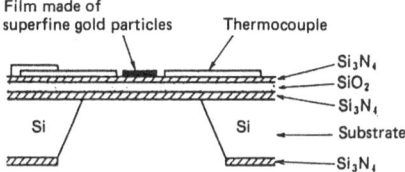

Fig. 6.42 Cross-section of infrared sensor using a film made from superfine gold particles (1).

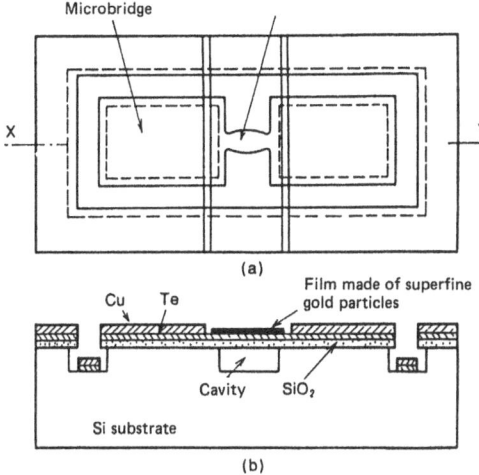

Fig. 6.43 Cross-section of infrared sensor using a film made of superfine gold particles (2).

When the hot junction of the thermocouple is set up on the thin film adjacent to the superfine gold particle film, 95% of the condensed infrared rays are absorbed by the film, which becomes hot, so the temperature difference between the hot and cold junctions can be read as thermoelectromotive force. The sensitivity is 95 V/W in air, and 799 V/W in a vacuum; the response speed is 30 ms [68].

Figure 6.43 shows a bolometer-type infrared sensor. The SiO_2 film is 1.5 μm thick, and the Te film formed by vapor deposition in a vacuum is 100 nm thick, with a temperature coefficient of $1.2 \times 10^{-2}/°C$. The thin Te film is the bolometer element; the thermoelectromotive force is 400 μV/°C; the sensitivity is 130 V/W, and the response speed is 200 μs.

References

1. Y. Takeda: Kogyou Zairyou, 31, 7, p. 24 (1983) (in Japanese)
2. M.K. Wu, J.R. Ashburn, C.J. Torng, P.H. Hor, R.L. Meng, L. Gao, Z.J. Huang, Y.Q. Wang, C.W. Chu: Phys. Rev. Lett., 58, p. 908 (1987)
3. K. Niwa: Kougyou Zairyou, 35, 12, p. 98 (1987) (in Japanese)
4. Y. Takeda: Ceramics, 19, 6, p. 489 (1984) (in Japanese)
5. Inui, Nakamura, Kitahara, Shinohara, Hirai: 2nd Microelectronics Symposium papers, p. 181 (1987) (in Japanese)
6. M. Ono: Electronic Ceramics, 17, 3, p. 21 (1986) (in Japanese)
7. T. Ido: Electronic Ceramics, 16, Winter, p. 49 (1985) (in Japanese)
8. M. Kiyama, T. Takada, N. Nagai, N. Horiishi: Fourth International Conference on Ferrites, San Francisco, Extended Abstracts, p. 226 (1984) (in Japanese)
9. Sugimoto, Arai: Electronic Ceramics, 17, 3, p. 52 (1986) (in Japanese)
10. Yokoyama, Koike, Suzuki: Toshiba Review, 40, 13, p. 1111 (1985) (in Japanese)
11. Imamura, Itou, Fujiki, Kubota: Toshiba Review, 40, 13, p. 1115 (1985) (in Japanese)
12. S.S. Papell, US Patent No. 3215572 (1965). Satou, Higuchi, Shimoiizaka: Chemical Society of Japan, 19, Nenkai Papers I, p. 293 (1966) (in Japanese). G.W. Reimers, S.E. Khalafalla: Bureau of Mines Technical Progress Report, p. 59 (1972). T. Wakao: 6th Ferrite Summer Seminar Abstracts, Funtai Funmatsu Yakin Kyoukai, p. 35 (1975) (in Japanese). R.E. Rosensweig, J.W. Nestor, R.M. Timmins: A.I.Ch.E., Chem. E. Symposium, 5, p. 104 (1965)
13. I. Nakatani: Cho-biryushi no Jitsuyouka Gijutsu, p. 135, CMC (1984) (in Japanese)
14. J.R. Thomas: J. Appl. Phys., 37, p. 2914 (1966)
15. A.E. Berkowitz, J.L. Walter, K.F. Wall: Phy. Rev. Lett., 46, 22, p. 1484 (1981)
16. Seki, Satou, Nakazato: Electronic Ceramics, 17, Winter, p. 64 (1985) (in Japanese)
17. Chemical Society of Japan (ed.): Cho-biryushi Gijutsu, p. 167, Gakkai Shuppan Center (1985) (in Japanese)
18. A. Makijima: Kinousei Glass Nyuumon, p. 9, Agne Gijutsu Center (1984) (in Japanese)
19. S. Saka: Hyoumen, 19, 8, p. 430 (1981) (in Japanese)
20. N. Ichinose: Electronics Ceramics, 16, 5, p. 36 (1985) (in Japanese)
21. Honda, Yuge, Ishizaki, Watanabe: Toshiba Review, 39, 3, p. 192 (1984) (in Japanese)
22. M. Fukuhara: Yougyou Kyoukai Ensetsukai Superfine Ceramics Papers, p. 53 (1985) (in Japanese)
23. E. Orowan: Rep. Prog. Phys., 12, p. 185 (1949)
24. Y. Ando, M. Ohkohchi: J. Cryst. Growth, 60, p. 147 (1982)
25. H. Suzuki (ed.): Kouon Ceramic Zairyou, p. 84, Nikkan Kogyo Shinbunsha (1985)
26. Chemical Society of Japan (ed.): Cho-biryushi Gijutsu, p. 179, Gakkai Shuppan Center (1985) (in Japanese)
27. Chemical Society of Japan (ed.): Cho-biryishi Gijutsu, p. 185, Gakkai Shuppan Center (1985) (in Japanese)
28. T.K. Gupta: Science of Sintering, 10, p. 205 (1978)
29. Katou, K. Yamaguchi: New Ceramic Funtai Handbook, p. 144, Science Forum Sha (1983)
30. W.H. Rhodes: J. Am. Ceram. Soc., 64, 1, p. 19 (1981)
31. Ozaki, Suzuki: Shouketsuyou Electronics Ceramics, 16, 7, p. 33 (1985) (in Japanese)
32. M. Ishii: FC Report, 1, 7, p. 10 (1983) (in Japanese)
33. Katou, Yoshimoto: Electronics Ceramics, 16, 11, p. 41 (1985) (in Japanese)
34. Tashiro, Sasaki, Tsuji, Igarashi, Okazaki: 6th Kyouyuudentai Ouyou Kaigi Papers, p. 90 (1987) (in Japanese)
35. Chemical Society of Japan (ed.): Cho-biryushi Gijutsu, p. 193, Gakkai Shuppan Center (1985) (in Japanese)
36. M. Mitarai: Ceramics, 19, 6, p. 495 (1984) (in Japanese)
37. D.W McKee: J. Phys. Chem., 67, p. 841 (1963)
38. Tokoro, Hori, Yashi, Uchijima, Yoneda: Nihon Kagakukaishi, 12, p. 1646 (1979) (in Japanese)

39. Noda, Shinoda, Saitou: Nihon Kagakukaishi, p. 1017 (1984) (in Japanese)
40. Hirai, Tojima: Koubunshi Satai Shokubai, Gakkai Shuppan Center, p. 56 (1982) (in Japanese)
41. Chemical Society of Japan (ed.): Cho-biryushi Gijutsu, p. 197, Gakkai Shuppan Center (1985) (in Japanese)
42. A. Ueno: Hyoumen, 22, p. 18 (1984) (in Japanese)
43. Uchijima, Kunimori: Hyoumen Kagaku, 5, p. 36 (1984) (in Japanese)
44. D.J.C. Yates, L.L. Murrell, and E.B. Prestridge: J. Catal., 57, p. 41 (1979)
45. P. Gallezot: Catal. Rev., 20, 1, p. 121 (1979)
46. T. Okuhara, T. Kimura, K. Kobayashi, M. Misono, Y. Yoneda: Bull. Chem. Soc. Jpn., 57, p. 938 (1984)
47. M. Ichikawa: J. Catal., 59, p. 67 (1979)
48. K.J. Klabunde, Y. Imizu: J. Am. Chem. Soc., 106, p. 2721 (1984)
49. Takasu, Kasahara, Matsuda, Toyoshima: Nihon Kagakukaishi, p. 1011 (1984) (in Japanese)
50. J.M. Tri, J. Massardier, P. Gallezot, B. Imelik: J. Catal., 85, p. 244 (1984)
51. Y. Iwasawa, M. Yamada: J. Chem. Soc. Chem. Commun., p. 675 (1985)
52. A. Ueno: Shokubai, 27, p. 71 (1985) (in Japanese)
53. A. Ueno, H. Suzuki, Y. Kotera: J. Chem. Soc. Faraday Trans. I, 27, p. 71 (1985)
54. Mizukami, Niwa, Isoyama, Tsuchiya, Shimizu, Imamura: Shokubai, 26, p. 402 (1984) (in Japanese)
55. K. Matsuo, K.J. Klabunde: J. Catal., 73, p. 216 (1982)
56. A. Ueno, H. Suzuki, Y. Kotera: J. Chem. Soc. Faraday Trans. I, 79, p. 127 (1983)
57. T. Okuhara, K. Kobayashi, T. Kimura, M. Misono, Y. Yoneda: J. Chem. Soc. Chem. Commun., p. 1114 (1981)
58. D. Commereuc, Y. Chauvin, F. Hugues, J.M. Basset, D. Olivier: J. Chem. Soc. Chem. Commun., p. 154 (1980)
59. T. Seiyama, A. Kato, K. Fujiishi, M. Nagatani: Anal. Chem., 34, p. 1502 (1962)
60. N. Yamazoe: Denki Kagaku Kyoukaishi, 50, p. 29 (1982) (in Japanese)
61. Matsuoka, Nakatani: Electronics Ceramics, 15, Summer, p. 19 (1984) (in Japanese)
62. Y. Nakatani, M. Sakai, M. Matsuoka: Proceedings of the International Meeting on Chemical Sensors, Kodansha-Elsevier, p. 147 (1983)
63. A. Abe: Electronics Ceramics, 16, 7, p. 22 (1985) (in Japanese)
64. H. Ogawa, A. Abe, M. Nishikawa, S. Hayakawa: J. Electrochem. Soc., 128, p. 685 (1981)
65. H. Ogawa, N. Nishikawa, A. Ade: J. Appl. Phys., 53, p. 4448 (1982)
66. A. Abe: Cho-biryushi Kotei Butsuri Bessatsu Tokushugou, p. 131 (1984) (in Japanese)
67. A. Abe: National Tech. Rep., 22, p. 853 (1976) (in Japanese)
68. Inoue, Kimura, Sekimukou: Thermopile Sekigaisen Sensors, Electronic Materials, 19, 9, p. 98 (1980) (in Japanese)

Additional References

69. J. Livage, M. Henry, J.P. Jolivet, C. Sanchez: "Chemical synthesis of fine powders", MRS Bulletin, pp. 18–25 (Jan. 1990)
70. M. Ozaki, "Preparation and properties of well-defined magnetic particles", MRS Bulletin, pp. 35–40 (Dec. 1989)
71. T. Sugimoto; "Preparation and characterization of monodispersed colloidal particles", MRS Bulletin, pp. 23–28 (Dec. 1989)
72. F. Fievet, J.P. Lagier, M. Figlarz: "Preparing monodisperse metal powders in micrometer and submicrometer sizes by the polyol process", MRS Bulletin, pp. 29–34 (Dec. 1989)
73. A.J.I. Ward, S.E. Friberg: "Preparing narrow size distribution particles from amphiphilic association structures", MRS Bulletin, pp. 41–46 (Dec. 1989)
74. E.J. Davis, M.F. Buehler: "Chemical reactions with single microparticles", MRS Bulletin, pp. 26–33 (Jan. 1990)

Subject Index